Mass Spectrometry in Life Sciences and Clinical Laboratory

Edited by Goran Mitulović

Published in London, United Kingdom

IntechOpen

Supporting open minds since 2005

Mass Spectrometry in Life Sciences and Clinical Laboratory
http://dx.doi.org/10.5772/intechopen.91573
Edited by Goran Mitulović

Contributors
Katherine A. Stumpo, Jadranka Miletić Vukajlović, Tanja Panić-Janković, Simone König, Anna Walke, Eric Suero Molina, Walter Stummer, Petr Vaňhara, Josef Havel, Aleš Hampl, Lukáš Moráň, Lukáš Pečinka, Tiziana Pivetta, Jose Elias Conde, Eladia Peña-Mendez, Sebastiano Masuri, Volodymyr Porokh, Irena Dapic, Jakub Faktor, David R. Goodlett, Ahmad Amini, Torgny Rundlöf, Henrik Lodén, Johan Carlsson, Karin Björk, Torbjörn Arvidsson, Martin Laven, Ezra Mulugeta, Iréne Agerkvist, Anette Perolari, Goran Mitulović

First published in London, United Kingdom, 2021 by IntechOpen
IntechOpen is the global imprint of INTECHOPEN LIMITED, registered in England and Wales, registration number: 11086078, 5 Princes Gate Court, London, SW7 2QJ, United Kingdom
Printed in Croatia

British Library Cataloguing-in-Publication Data
A catalogue record for this book is available from the British Library

Additional hard and PDF copies can be obtained from orders@intechopen.com

Mass Spectrometry in Life Sciences and Clinical Laboratory
Edited by Goran Mitulović
p. cm.
Print ISBN 978-1-83881-861-6
Online ISBN 978-1-83881-862-3
eBook (PDF) ISBN 978-1-83881-863-0

For EU product safety concerns:
IN TECH d.o.o., Prolaz Marije Krucifikse Kozulić 3, 51000 Rijeka, Croatia, info@intechopen.com or visit our website at intechopen.com.

Meet the editor

Dr. Goran Mitulović (ORCID 0000-0003-1964-3965) is the head of the Proteomics Core Facility, Medical University of Vienna. He has more than twenty years of experience developing hardware and analytical methods for multidimensional nano HPLC-MS separation and detection methods for proteins and peptides. He conducted his research with LC Packings/Dionex in Amsterdam; IMP and IMBA in Vienna; and the Medical University of Vienna. His research interests include mechanistic chromatographic separation of peptides and proteins on different stationary phases (polar, ion-exchange, hydrophobic, and hydrophilic phases), improvement of mass spectrometric response based on the optimized composition of the chromatographic mobile phase, and development of clinical proteomics methods and approaches for use in diagnostics.

Contents

Preface

Mass spectrometry (MS) is one of the most used analytical tools for analyzing complex organic and inorganic samples. Analyses of biological samples in life sciences, medical sciences, clinical laboratories, food processing, research, and quality control in the pharmaceutical industry are routinely performed using different MS approaches. Furthermore, MS is applied for controlling the origin of food and identifying mislabeled products and toxins in foods [1–4]. Identification of illicit substances and drugs of abuse relies heavily on MS, and its use is widely reported [5–11]. Technological innovations, such as improved ion transfer, improved ionization efficiency, and mass resolution of mass spectrometers, have significantly improved the data generated and widened the field of use.

A significant number of analytical methods for MS in clinical laboratories have been developed in the recent past. Development of these methods has rapidly increased as the COVID-19 pandemic holds humankind in its grip, and techniques for fast high-performance liquid chromatography-mass spectrometry (HPLC-MS) have been developed and applied.

In the chapter "Mass Spectrometry in Clinical Laboratories," Miletić Vukajlović and Panić-Janković discuss the development of MS methods for newborn screening and detection of modifications in the metabolome of amino acids that can lead to severe health conditions. Identifying such changes has helped develop more effective treatment methods. The authors further discuss the detection and analysis of antibiotics in clinical samples, which is an important topic considering the growing resistance of bacteria against the antibiotics currently in use.

In the chapter "Identification of Peptides and Proteins in Illegally Distributed Products by MALDI-TOF-MS," Amini et al. address and discuss using MAlDI MS to trace illegal products. The market for illegal medical products, pharmaceuticals, and food is an ever-growing one, and counterfeit products are a severe threat to human health. Proteins and peptides can be identified and characterized quickly by applying MALDI, and illicit products can be traced and eliminated from the market.

Research and analysis of stem cells has given hope to many regarding the treatment of rare diseases and significant health conditions. Vaňhara et al. discuss the mass spectrometry methods for analyzing and characterizing the single intact embryonic stem cell before being used for a patient's treatment In the chapter "Intact Cell Mass Spectrometry for Embryonic Stem Cell Biotyping," Vaňhara et al. discuss MS methods for analyzing and characterizing a single, intact embryonic stem cell before using it to treat patients. MALDI-TOF MS, based on the principle of mass spectrometric "biotyping" of bacteria, is currently being used to analyze single cells. Mass spectra generated by MALDI contain information that helps to reveal the cell's immunophenotype and activation state. Results from MS allow for the identification of changes and posttranslational modifications associated with the molecular phenotype of the applied cell lines.

König et al. describes HPLC hyphenated to M.S. for Identification of Protoporphyrin IX (PPIX) that can serve as a potential biomarker for cancer screening in neurological cancer, especially the neuroblastoma In the chapter "Protoporphyrin IX Analysis from Blood and Serum in the Context of Neurosurgery of Glioblastoma" König et al. describe HPLC for identifying protoporphyrin IX (PPIX), which is a potential biomarker for cancer screening in neurological cancer, especially neuroblastoma. The benefit of using MS to analyze a complex biological sample is shown in its ability to detect different forms of metal-conjugated PPIX based on fragmentation mass spectra. The methods described also show that MS has a significant advantage over the traditionally used fluorescence-based methods.

One of the fields with the most dynamic growth in the clinical application of MS is mass spectrometry imaging (MSI). The improvement of laser technology with better and more sensitive detectors has shown the enormous potential of this technology. MSI in the clinical field is still overwhelmingly used for digital pathology, particularly for brain imaging. In the chapter "Mass Spectrometry Imaging of Neurotransmitters," Stumpo describes the use of MSI for imaging neurotransmitters, which have broad and significant implications due to their involvement in the complex biological processes that occur in the brain.

The most advanced analytical instrumentation cannot fix the mistakes made during sample preparation for HPLC-MS analysis. Dapić et al. address this critical topic In the chapter "Trends in Sample Preparation for Proteome Analysis," Dapić et al. addresses this critical topic. Depending on the analysis to be performed, different types of sample preparation are utilized. This step is the key to successful research in proteomics. However, there is still no consensus in the proteomics and clinical community on how to treat the biological sample and which set-up should be used to prepare proteins from clinical samples. This chapter focuses on sample preparation for bottom-up proteome profiling using MS and, especially, for preparing tissue samples from fresh frozen and formalin-fixed paraffin-embedded material.

MS is a powerful analytical technique that still has not developed its full potential for use in clinical laboratories, although its use is growing. Different MS-based devices have been approved for screening newborns, identifying microbes and fungus in cultures of human cells, and detecting and measuring the concentrations of drugs (therapeutic and illicit) in body fluids. The development of large and integrated HPLC-MS systems for detecting and measuring peptides or proteins in clinical laboratories is still yet to happen. However, MS has already gained access to all areas of medical research and diagnostics.

Goran Mitulović
Medical University of Vienna,
Austria

References

[1] Colaco CMG, Basile K, Draper J, Ferguson PE. Fulminant Bacillus cereus food poisoning with fatal multi-organ failure. BMJ Case Rep. 2021;14(1).

[2] Shannon M, Ratnasekhar CH, McGrath TF, Kapil AP, Elliott CT. A two-tiered system of analysis to tackle rice fraud: The Indian Basmati study. Talanta. 2021;225:122038.

[3] Wu Y, Ye J, Xuan Z, Li L, Wang H, Wang S, et al. Development and validation of a rapid and efficient method for simultaneous determination of mycotoxins in coix seed using one-step extraction and UHPLC-HRMS. Food Addit Contam Part A Chem Anal Control Expo Risk Assess. 2021;38(1):148-59.

[4] Gatmaitan AN, Lin JQ, Zhang J, Eberlin LS. Rapid Analysis and Authentication of Meat Using the MasSpec Pen Technology. J Agric Food Chem. 2021;69(11):3527-36.

[5] Wallace B, Hills R, Rothwell J, Kumar D, Garber I, van Roode T, et al. Implementing an Integrated Multi-Technology Platform for Drug Checking: Social, Scientific, and Technological Considerations. Drug Test Anal. 2021.

[6] Li L, Quintero X. Rapid orthogonal screening of forensic drugs using a compact UHPLC-PDA/SQD; a complement to GC-MS. Forensic Sci Int. 2021;318:110565.

[7] de Araujo WR, Cardoso TMG, da Rocha RG, Santana MHP, Muñoz RAA, Richter EM, et al. Portable analytical platforms for forensic chemistry: A review. Analytica chimica acta. 2018;1034:1-21.

[8] Strano-Rossi S, Odoardi S, Fisichella M, Anzillotti L, Gottardo R, Tagliaro F. Screening for new psychoactive substances in hair by ultrahigh performance liquid chromatography–electrospray ionization tandem mass spectrometry. Journal of Chromatography A. 2014;1372(0):145-56.

[9] Ostermann KM, Luf A, Lutsch NM, Dieplinger R, Mechtler TP, Metz TF, et al. MALDI Orbitrap mass spectrometry for fast and simplified analysis of novel street and designer drugs. Clinica Chimica Acta. 2014;433(0):254-8.

[10] Hendricks PI, Dalgleish JK, Shelley JT, Kirleis MA, McNicholas MT, Li L, et al. Autonomous in situ analysis and real-time chemical detection using a backpack miniature mass spectrometer: concept, instrumentation development, and performance. Anal Chem. 2014;86(6):2900-8

[11] Baker DR, Kasprzyk-Hordern B. Multi-residue analysis of drugs of abuse in wastewater and surface water by solid-phase extraction and liquid chromatography-positive electrospray ionisation tandem mass spectrometry. Journal of Chromatography A. 2011;1218(12):1620-31.

Section 1

Introduction

Chapter 1

Introductory Chapter: A Tool for Aided Advanced Diagnostics and Deep View into Biological Sample

Goran Mitulović

1. Introduction

The mass spectrometry (MS) is a technology enabling the measurement and analysis of multiple analytes with very high sensitivity and selectivity. The MS "use for purpose" includes the analysis of proteins and their modifications using either the top-down or the bottom-up approach, the analysis of multiple parameters for toxicology research, targeted analysis of active pharmaceutical compounds in patients' samples.

Recent years has brought a significant improvement of technology for ion-trap, the time-of-fligt, and triple quadrupole instruments resulting with a surge in applications for the medical, biological, and inorganic field. One of the greatest challenges for the MS was how to solve the problem of the insufficient ion transfer from the ionization source through different mass filters and to the detector. This challenge can be considered essentially solved due to changes in the design of ion-sources and subsequent mass filters such as ion funnels of different design, which enabled hugely improved ion focusing and ion transfer. However, there is still enough space to further improve the design of the interface to enable handling of larger ion currents generated by more powerful and more intense electrospray (ESI) ion sources. That would make bulky, expensive, and complex pumping systems obsolete. Actually, the greatest challenge of improving the MS is the ionization step, particularly when MS is used in combination with the liquid chromatography (LC). The combination of LC and MS (LC–MS) has advanced to the main workhorse in many laboratories, especially in biotechnology and medical laboratories. Since the low LC flow rate significantly increases the efficiency, the sensitivity, and the stability of the ESI, the use of separation columns with small inner diameters is recommended for the hyphenation of the two technologies. In cases with very low flow rates, e.g. 5–50 nl/min, a complete ionization of a substance can be achieved. On the other side, the use of such a low flow rates causes problems with flow's stability, reproducible flow gradient mixing, and stable ESI performance.

2. Clinical laboratory and mass spectrometry

The laboratory medicine along with medical imaging procedures is one of the pillars of the modern diagnostics. The laboratory medicine has significantly benefitted from technical and technological development of analytical chemistry, miniaturization of instruments, and optimization of analytical methods.

Colorimeter and spectromphoter were the first modern analytical instruments to be used in a clinical laboratory. Since the time of their introduction, the art of

 IntechOpen

performing analyses and tests has significantly changed. In the 1950's additional developments were made and, in 1957 and 1959, respectively, the autoanalyzer, which is the precursor to the modern analytical systems, and the first RIA (immunoassay) for analysis of insulin were introduced. The introduction of RIA, which is still widely used, for insulin has significantly improved and changed the art of measuring a large of compounds.

The introduction of the mass spectrometry (MS) into the clinical laboratory had had the same revolutionary impact as the previously mentioned methods. A brief search for "mass spectrometry" and "clinical" from 1950 until 1970 results with only 9 publications! During the next 20 years, until 1990, the number of publications referring to the use of MS for different analyses in clinical laboratory jumped to 798! From the early 1990's until today, the steep rise of MS methods and approaches for analyses in clinical laboratory has steeply raised with more than 5000 publications from January 1st, 2020 – February 1st, 2021.

The significant rise in use of different MS approaches for clinical analyses correlates with improvements in ionization technologies, miniaturization of separation systems, notably of chromatographic systems (HPLC) [1], the significant and exciting improvements in sample preparation even of a single cell [2], and bioinformatic analysis. The mass spectrometry is applied for both "classical" clinical laboratory analyses and for analyzing samples for personalized and precision medicine. Undoubtly, the approaches and methods describe in current book are only a small part of possible applications.

In clinical laboratories, the analysis of clinical samples and monitoring levels of active compounds and their metabolites in e.g. patients' blood and urine samples are the main application fields. It is possible to perform specific detection of target analytes by applying MRM/SRM (multiple-reaction monitoring/selected-reaction monitoring) or SIM (single-ion monitoring) thus significantly enhancing the selectivity and sensitivity of the analytical method and provide targeted and highly specific analytical approach.

3. Separation approaches

Analytes of interest in complex biological samples must be separated prior to mass spectrometric detection and analysis and chromatography is the most widely used separation method for biological samples prior to MS. A number of Companies, e.g. Chromsystems (https://www.chromsystems.com/), ThermoFisher Scientific (https://www.thermofisher.com/at/en/home/clinical/diagnostic-testing/clinical-chemistry-drug-toxicology-testing/therapeutic-drug-monitoring.html), Biocrates (https://biocrates.com/) or BioRad (www.bio-rad.com) developed analytical systems for a broad range of analyses of important clinical parameters. Integrated HPLC–MS clinical systems were also developed but they never really did find broad acceptance. Sciex introduced the Topaz® system (https://www.businesswire.com/news/home/20170731005050/en/First-Fully-Integrated-LC-MS-System-for-Clinical-Diagnostics-Announce-by-SCIEX) and Thermo Fisher introduced the Cascadion SM lab analyzer (https://www.labmedica.com/aacc-2017/articles/294770271/worlds-first-fully-integrated-lc-ms-ms-clinical-analyzer-unveiled.html) in order to lower the barriers of many laboratories to adopt LC–MS.

The hyphenation of chromatography and mass spectrometry has its primary values in relatively fast detection and analysis of multiple analytes in a single sample with high sensitivity and high selectivity - the key challenge and requirement to detect and quantify low-concentration analytes. Currently, the most widely used separation columns for the HPLC–MS in a clinical laboratory have an inner diameter of 2 mm. The quality of electrospray is highly dependable on separation conditions, i.e. mobile phase, presence or absence of salts, flow speed, column's inner diameter, etc. In proteomics, the use of separation columns with 50 μm or 75 μm ID is state-of-the-art; however, the columns operated at a low flow rate of several hundreds of nanoliters/minute are still rare in clinical analysis although they can provide a significant increase of sensitivity. Currently, the use of nanoflow separation still cannot cope with the demand for high sample throughput and robustness in clinical applications. Currently, the closest compromise between sensitivity and throughput is the use of the microbore and the capillary columns of 300 μm – 500 μm and 1 mm – 2 mm inner diameter.

A new and exciting application of mass spectrometry in the clinical environment is the use of "live-MS" during surgical operations. Further development of this approach will revolutionize the diagnostics and help surgeons in extracting e.g. tumors with higher accuracy and better prognosis for the patient following the surgery [3–5].

In addition of analyzing small molecules in a targeted approach, the mass spectrometry can be applied in a clinical laboratory for a more widely screening approaches, e.g. screening of the human metabolome. The metabolome shed a light on our biological life story, reviling changes and processes that happened due to our genetics and due to the influence of the environment and the lifestyle. The measurement, detection, and analysis of metabolites is a step towards profiling an individuals' metabolic profile at any given time. That information can help understanding and, eventually, predicting the impact of environmental factors on the health. Therefore, metabolomics, in combination with other omics methods is a potent addition to developing personalized medical approaches.

A number of analytical methods for mass spectrometry in clinical laboratories were developed during the recent past, and the development gains additional momentum as the CoVID-19 pandemics holds the mankind in grip and methods for fast HPLC–MS have been developed and applied [6–15].

4. Clinical proteomics

The use of MS technology for measurement and analysis of clinically important peptides and protein biomarkers will definitively increase with further improving MS technology. It is clear that personalized medicine will become the major field of further development of targeted therapy and MS will be one of the major players in identifying both therapeutically targets and therapeutical agents. It is with great certainty that MS applications for evaluation of protein and peptide-based, or even the mRNA-based therapeutics, will play a crucial part for the quality control of the therapeutics, evaluating drug efficacy, or investigating therapeutic response.

Another issue is the miniaturization of MS and LC–MS systems that can be used portable systems or be applied in smaller field-laboratories. Certain advances were already achieved on developing such devises [16–23] that are being used in a number of analyses.

To conclude, mass spectrometry is a powerful analytical technology that still has not developed her full potential for use in clinical laboratory although its' use is growing. Different mass spectrometry-based devices have been approved for screening newborns, identifying microbes and fungus in cultures of human cells or for detecting measuring the concentrations of drugs (therapeutic and illicit) in body fluids. The development of large and integrated HPLC–MS systems for detecting and measuring peptides or proteins in clinical laboratories is still waiting to happen but mass spectrometry has already gained access to all areas of medical research and diagnostics.

Author details

Goran Mitulović
Proteomics Core Facility, Medical University of Vienna, Vienna, Austria

*Address all correspondence to: goran.mitulovic@meduniwien.ac.at

References

[1] Chervet JP, Ursem M, Salzmann JP. Instrumental Requirements for Nanoscale Liquid Chromatography. Analytical Chemistry. 1996;68(9):1507-12.

[2] Bensen RC, Standke SJ, Colby DH, Kothapalli NR, Le-McClain AT, Patten MA, et al. Single Cell Mass Spectrometry Quantification of Anticancer Drugs: Proof of Concept in Cancer Patients. ACS pharmacology & translational science. 2021;4(1):96-100.

[3] Alexander J, Gildea L, Balog J, Speller A, McKenzie J, Muirhead L, et al. A novel methodology for in vivo endoscopic phenotyping of colorectal cancer based on real-time analysis of the mucosal lipidome: a prospective observational study of the iKnife. Surgical endoscopy. 2017;31(3):1361-70.

[4] St John ER, Balog J, McKenzie JS, Rossi M, Covington A, Muirhead L, et al. Rapid evaporative ionisation mass spectrometry of electrosurgical vapours for the identification of breast pathology: towards an intelligent knife for breast cancer surgery. Breast cancer research : BCR. 2017;19(1):59.

[5] Luptakova D, Pluhacek T, Palyzova A, Prichystal J, Balog J, Lemr K, et al. Meet interesting abbreviations in clinical mass spectrometry: from compound classification by REIMS to multimodal and mass spectrometry imaging (MSI). Acta virologica. 2017;61(3):353-60.

[6] Páez-Franco JC, Torres-Ruiz J, Sosa-Hernández VA, Cervantes-Díaz R, Romero-Ramírez S, Pérez-Fragoso A, et al. Metabolomics analysis reveals a modified amino acid metabolism that correlates with altered oxygen homeostasis in COVID-19 patients. Scientific reports. 2021;11(1):6350.

[7] Pearson LA, Green CJ, Lin D, Petit AP, Gray DW, Cowling VH, et al. Development of a High-Throughput Screening Assay to Identify Inhibitors of the SARS-CoV-2 Guanine-N7-Methyltransferase Using RapidFire Mass Spectrometry. SLAS Discov. 2021:24725552211000652.

[8] Xie L, Wang Y, Yin H, Li J, Xu Z, Sun Z, et al. Identification of the absorbed ingredients and metabolites in rats after an intravenous administration of Tanreqing injection using high-performance liquid chromatography coupled with quadrupole time-of-flight mass spectrometry. Journal of separation science. 2021.

[9] St-Germain JR, Astori A, Raught B. A SARS-CoV-2 Peptide Spectral Library Enables Rapid, Sensitive Identification of Virus Peptides in Complex Biological Samples. Journal of proteome research. 2021.

[10] Sok V, Marzan F, Gingrich D, Aweeka F, Huang L. Development and validation of an LC-MS/MS method for determination of hydroxychloroquine, its two metabolites, and azithromycin in EDTA-treated human plasma. PloS one. 2021;16(3):e0247356.

[11] Yan L, Yi J, Huang C, Zhang J, Fu S, Li Z, et al. Rapid Detection of COVID-19 Using MALDI-TOF-Based Serum Peptidome Profiling. Anal Chem. 2021.

[12] Shi D, Yan R, Lv L, Jiang H, Lu Y, Sheng J, et al. The serum metabolome of COVID-19 patients is distinctive and predictive. Metabolism. 2021;118:154739.

[13] Xu J, Zhou M, Luo P, Yin Z, Wang S, Liao T, et al. Plasma metabolomic profiling of patients recovered from COVID-19 with pulmonary sequelae 3 months after discharge. Clin Infect Dis. 2021.

[14] Malla TR, Tumber A, John T, Brewitz L, Strain-Damerell C, Owen CD, et al. Mass spectrometry reveals potential of β-lactams as SARS-CoV-2 M(pro) inhibitors. Chemical communications (Cambridge, England). 2021;57(12):1430-3.

[15] Sørensen LK, Petersen A, Granfeldt A, Simonsen U, Hasselstrøm JB. A validated UHPLC-MS/MS method for rapid determination of senicapoc in plasma samples. Journal of pharmaceutical and biomedical analysis. 2021;197:113956.

[16] Meisenbichler C, Kluibenschedl F, Müller T. A 3-in-1 Hand-Held Ambient Mass Spectrometry Interface for Identification and 2D Localization of Chemicals on Surfaces. Anal Chem. 2020;92(21):14314-8.

[17] Wang X, Zhou X, Ouyang Z. Direct Analysis of Nonvolatile Chemical Compounds on Surfaces Using a Hand-Held Mass Spectrometer with Synchronized Discharge Ionization Function. Anal Chem. 2016;88(1):826-31.

[18] Hendricks PI, Dalgleish JK, Shelley JT, Kirleis MA, McNicholas MT, Li L, et al. Autonomous in situ analysis and real-time chemical detection using a backpack miniature mass spectrometer: concept, instrumentation development, and performance. Anal Chem. 2014;86(6):2900-8.

[19] Chen TC, Ouyang Z. Synchronized discharge ionization for analysis of volatile organic compounds using a hand-held ion trap mass spectrometer. Anal Chem. 2013;85(3):1767-72.

[20] Dunn JD, Gryniewicz-Ruzicka CM, Kauffman JF, Westenberger BJ, Buhse LF. Using a portable ion mobility spectrometer to screen dietary supplements for sibutramine. Journal of pharmaceutical and biomedical analysis. 2011;54(3):469-74.

[21] Sanders NL, Sokol E, Perry RH, Huang G, Noll RJ, Duncan JS, et al. Hand-held mass spectrometer for environmentally relevant analytes using a variety of sampling and ionization methods. Eur J Mass Spectrom (Chichester). 2010;16(1):11-20.

[22] Gao L, Sugiarto A, Harper JD, Cooks RG, Ouyang Z. Design and characterization of a multisource hand-held tandem mass spectrometer. Anal Chem. 2008;80(19):7198-205.

[23] Keil A, Talaty N, Janfelt C, Noll RJ, Gao L, Ouyang Z, et al. Ambient mass spectrometry with a handheld mass spectrometer at high pressure. Anal Chem. 2007;79(20):7734-9.

Section 2

Sample Preparation

Chapter 2

Trends in Sample Preparation for Proteome Analysis

Jakub Faktor, David R. Goodlett and Irena Dapic

Abstract

Sample preparation is a key step in proteomics, however there is no consensus in the community about the standard method for preparation of proteins from clinical samples like tissues or biofluids. In this chapter, we will discuss some important steps in sample preparation used for bottom-up proteome profiling with mass spectrometry (MS). Specifically, tissues, which are an important source of biological information, are of interest because of their availability. Tissues are most often stored as fresh frozen (FF) or formalin-fixed paraffin-embedded (FFPE). While FF tissues are more readily available, paraffin embedding has historically been routinely used for tissue preservation. However, formaldehyde induced crosslinks during FFPE tissue preservation present a challenge to the protocols used for protein retrieval. Moreover, in our view, an important aspect to consider is also the amount of material available at the start of a protocol since this is directly related to the choice of protocol in order to minimize sample loss and maximize detection of peptides by MS. This "MS sensitivity" is of special importance when working with patient samples that are unique and often available in limited amounts making optimization of methods to analyze the proteins therein important given that their molecular information can be used in a patients' diagnosis and treatment.

Keywords: sample preparation, tissue, digestion, mass spectrometry, proteomics

1. Introduction

Proteomics is an important tool in the study of human biological material with the aim to extract knowledge that can improve a patients' treatment outcomes. Molecular information obtained from patient samples can be complementary to pathological observations all with the goal of faster and more accurate diagnosis, and subsequent treatment. Molecular analysis of tissue by proteomics can lead to disease classification and reveal underlying disease pathways that can further serve as a target for medical treatment.

Sample size and origin is an important aspect in sample preparation. Today, there are numerous sample preparation procedures existing which aim to improve sensitivity of detection or protein recovery from a sample. Release of proteins from native or artificial material is a crucial step in sample preparation and to improve protein recovery there are different additives such as detergents, chaotropes, buffers and salts added during the sample preparation that must be considered. Moreover, targeting special groups of proteins (e.g. membrane proteins), which are involved in key cellular functions and may be a target of pharmaceutical treatment, often represent a challenge in their isolation and analysis. Their amphipathic nature may require use of appropriate enrichment procedures all with the goal to achieve better detection.

Further, sample loss during most standard preparation procedures is inevitable, and it is even more accentuated when minute amounts of material are being processed. To minimize sample loss and thus increase sensitivity of the analysis at the MS step there have recently been several technologies developed. Specifically, improvement in technologies that allow detection of proteins down to a single cell have become available. Some of these technologies, such as nanoPOTS and microPOTS, have already been applied to human tissues. These new possibilities to analyse small regions of tissue samples with sufficient sensitivity is opening the door to many applications such as profiling of selected regions of a tumorous zone or detection of proteins from subcellular populations. These new applications aimed at working with 1 to 100s or 1000s of cells will likely have increasing importance in clinics, but only if they can be developed into routine and robust methods.

2. Tissue preservation

Human tissue samples are a valuable source of information for diagnostics, therefore a lot of effort has gone into best preservation methods that minimize changes that can occur over time in storage. For example, following clinical surgery tissues need to be stored according to the protocols that minimize chemical, enzymatic, mechanical or thermal degradation and protect their molecular content. Today, tissues are most often preserved as fresh frozen (FF) or formalin-fixed paraffin-embedded (FFPE) tissues.

2.1 Fresh frozen tissues

FF tissues are obtained usually with snap freezing of tissues where the temperature achieved is below −70°C, most often in dry ice or liquid nitrogen (**Figure 1B**). To minimize variability between the samples storage and thus to minimize potential effect on molecular structure and integrity of the tissue, the European Human Frozen Tumour tissue bank (TuBa-Frost) has standardized tissue preservation by freezing in 2006 [1, 2]. An important aspect in preservation of tissues by the FF method is prevention of formation of artefacts that might result in changes to the tissue structure and morphology. For example, ice crystals that can disrupt structures within the tissue may form as a consequence of the freezing procedure due to moisture present within the tissue [3]. An alternative to snap freezing is the optimal cutting temperature (OCT) compound, which is used for tissue embedding and contains polyvinyl alcohol, polyethylene glycol (PEG) and benzalonium chloride. The OCT substance preserves tissue and enables optimal microdissection of the tissue. However, where samples will later be analyzed by mass spectrometry (MS), OCT compounds must be removed prior to analysis. This is usually achieved by washing the tissue with a special grade of alcohol or Carnoy's fluid [4] or with the use of other protocols for sample purification.

2.2 Formalin-fixed paraffin-embedded tissues

An alternative to preservation of tissue by the FF process is the use of FFPE methods (**Figure 1A**), which are routinely used by pathologists around the globe to preserve tissue by embedding in paraffin. The FFPE process preserves tissues by chemical fixation most often in 10% of formalin and is followed later by embedding in paraffin to form a tissue block for subsequent slicing. The combination of formalin fixation with paraffin embedding allows for long term storage of tissues.

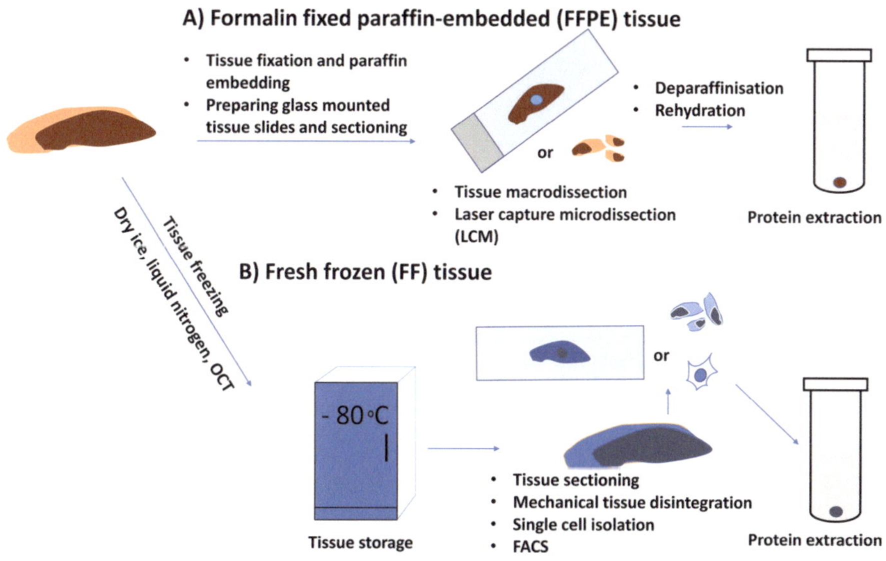

Figure 1.
Overview of tissue sample processing prior to proteomic experiments. A) Tissues can be preserved and stored for long periods of time by formalin fixation and paraffin embedding which has been routinely used for decades. Proteomic analysis of FFPE tissues can be accompanied with laser capture microdissection (LCM) which helps to retrieve regions of interest on tissue sections. Further steps involve isolation of the proteins from the sections in appropriate lysis buffers and further processing for protein analysis by MS. B) Tissues can be preserved by freezing and further stored at low temperatures. Tissue should be frozen as soon as possible after retrieval and tissue can be sectioned prior to proteomic sample preparation. Tissue sections are further prepared for protein isolation, but also single cell isolation protocols could be employed to retrieve limited cell subsets prior protein extraction.

Also, FFPE tissues are often used for histopathological studies, a routine process in examination of a patient's biopsies and clinical material [5]. Moreover, it is known that formalin leads to chemical modification of proteins in the fixed tissues causing cross-linking between proteins and modifications most often as methylation (+14 Da) as well as formation of methylene and methylol adducts to a lesser extent. As a consequence of formaldehyde induced modifications, the molecular weight or physicochemical properties of fixed proteins can be altered.

3. Preparation of the sample for bottom-up proteomics

Protein extraction and the subsequent preparation for LC–MS analysis represents one of the key steps in proteomics (**Figure 2**). While there have been numerous protocols reported, they have mainly focused on preparation from large amounts (i.e. micrograms to milligrams) of material, which limits their utility in the study of patient clinical samples. Notably, protein extraction from FFPE preserved tissues requires removal of formaldehyde-formed cross links, which is usually carried out by heating samples in a buffered solution at an elevated temperature (95°C or 100°C). The most common buffers used for protein extraction are ammonium bicarbonate, tris(hydroxymethyl)aminomethane (Tris), and Radioimmunoprecipitation assay (RIPA) buffer. Addition of detergents to the buffer composition (e.g. sodium dodecyl sulfate (SDS), sodium dodecyl cholate (SDC), RapiGest SF surfactant™ (Waters), PPS Silent Surfactant™ (Expedeon) have been routinely used to improve protein solubilization efficiency and thus enhance protein extraction. In addition to optimization of the extraction buffers many studies also optimized other parameters like incubation time of the extraction and/or addition of various proteases to improve protein coverage during subsequent LC–MS/MS analysis.

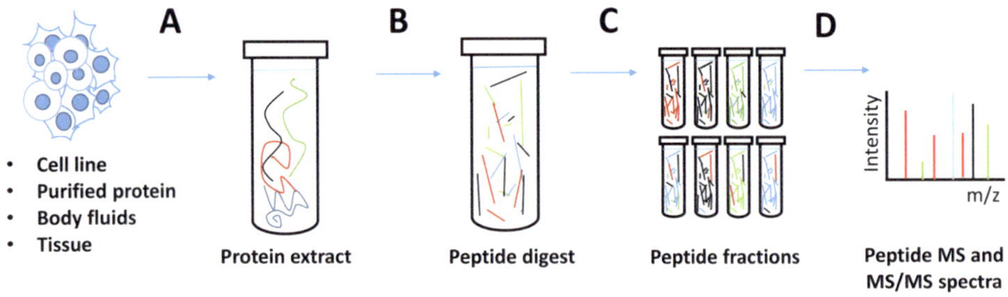

Figure 2.
Overview of sample preparation for bottom-up proteomic analysis by tandem mass spectrometry. A) Sample lysis: proteins are extracted from biological matrix in lysis buffer. Mechanical disintegration or sonification is used to homogenize rigid structures present within samples such as is common in mammalian tissue. B) Protein digestion: proteins are proteolytically digested into peptides, usually by the protease trypsin. C) Peptide fractionation: optionally, the complexity of the peptide sample is decreased by addition of fractionation steps orthogonal to methods used in the next step. D) Mass spectrometry analysis: desalted peptide samples are dissolved in an appropriate buffer and introduced into a tandem mass spectrometer. Most often reversed phase liquid chromatography separation is used in this final step to enable sequential introduction of peptides into the tandem mass spectrometer.

3.1 Detergents

Traditional detergents and chaotropes such as SDS and urea have been widely used for protein solubilization, however they are also well known to inhibit digestion at higher concentrations and are incompatible with reversed phase chromatography separation (RPLC) used to introduce samples for MS analysis. Therefore, their concentration must be kept low at the time of proteolysis in order to preserve the effectiveness of proteases used for protein digestion. Failure to do so often leads to incomplete protein solubilization and denaturation. Also, presence of detergents in the sample might interfere with later instrumental analysis, therefore there have been different purification methods developed for detergent removal to improve LC–MS outcome. The choice of the most effective procedure depends on the physicochemical properties of the detergent. Some of the procedures might include detergent removal on the basis of size exclusion (i.e. molecular weight cut-off filters) or with the use of spin columns containing appropriate resins for detergent removal. Moreover, heating of the sample in urea buffers often leads to covalent modification of proteins via carbamylation, which might affect peptide retention time during RPLC separation and if not accounted for will interfere with identification. In order to circumvent these problems caused by mass spectrometry incompatible detergents significant effort went into development of reagents that avoid these complications. To this end, acid labile detergents such as RapiGest SF surfactant™ (Waters) and PPS Silent Surfactant™ (Expedeon) were developed that could be easily removed after proteolysis by simple measures like decreasing the pH. For example, the MS compatible surfactant ProteaseMAX™ (Promega) surfactant enhances tryptic, chymotryptic and LysC digestion and then degrades during the course of a digestion reaction. Another compound, Invitrosol™ (Thermo Fisher Scientific) is a homogenous surfactant that does not impact tryptic digestion and elutes during RPLC in three peaks well separated from where peptides elute [6].

3.2 Sample digestion

Classical bottom-up proteomic sample preparation aims to turn protein extracts into peptides via a process of protein cleavage or digestion with proteases. Notably, proteins extracted from biological material tend to keep their native tertiary structure mostly held by non-covalent interactions of amino acid side groups [7].

It is thus essential to disrupt the tertiary structure and linearize the protein sequence to ease the accessibility of proteases to cleavage sites. Protein tertiary structure is frequently disrupted by chaotropic and denaturing reagents. Disulfide bonding contributes to tertiary structure as well via a covalent bond between cysteine side chain groups also termed an S-S bridge. Disulphide bonds are most often broken by use of reducing agents leaving free sulfhydryl groups available that allow the protein to unfold more fully. Dithiothreitol (DTT), tris (2-carboxyethyl) phosphine (TCEP), tris (3-hydroxypropyl) phosphine (THPP) and 2-mercapto-ethanol (2-ME) are the most commonly used reducing agents. Sulphur containing reagents such as 2-ME and DTT break the S-S bridge by thiol-disulfide exchange, while phosphorus containing reagents form a phosphine oxide as a result of disulphide bond reduction [8]. Reduction is commonly followed by free sulfhydryl group alkylation to prevent disulphide bond reformation. In this chemistry a free sulfhydryl group performs a nucleophilic attack on the alpha carbon of an alkylating reagent creating a covalent bond between the alkyl group and cysteine. There is a wide palette of alkylating reagents that may be used, but in proteomic sample preparation the most commonly used reagents include iodoacetamide, iodoacetic acid, N-ethylmaleimide (NEM) and S-methyl methanethiosulfonate. Covalent modification of a free sulfhydryl group leaves a mass tag on each cysteine that must be considered as a mass shift to cysteine during interpretation of peptide tandem mass spectra. Alkylated proteins are then further processed by proteolytic cleavage, to shorter segments; peptides, which are then easily detected in a bottom-up experiment carried out by LC–MS/MS analysis. As mentioned above peptides may be produced by enzymatic methods but also chemical methods that can be either specific or unspecific (**Table 1**). In both cases there are a variety of protocols available to digest proteins into peptides for mass spectrometry-based proteomic analysis.

Bottom-up proteomics frequently relies on proteolytic enzymes that digest a protein at specific sites. Having predictable digestion rules for a given protease results in a faster database search process that also makes it computationally less demanding and more accurate. Trypsin is the most common protease in bottom-up proteomics cleaving peptide bonds at the C-terminus of arginine and lysine when not followed by proline [16]. Notably, maintaining an optimal temperature of 37°C at a pH optimum between 7 and 8 in the presence of Ca^{2+} ions in the digestion buffer is important for the reaction to proceed efficiently [17]. The optimal enzyme to substrate ratio is also important and for trypsin this is often from 1:20 to 1:100 (w:w). In some instances LysC endoproteinase, which is isolated from *Lysobacter enzymogenes*, is often combined with trypsin to provide cleavage at lysine C-terminus. This combination of multiple enzymes is used to enhance peptide sequence coverage by producing overlapping peptides. The addition of chymotrypsin and pepsin produce the most orthogonal peptides to trypsin. Chymotrypsin is a serine protease which cleaves a peptide bond at the C-terminus of large hydrophobic side chain amino acids such as phenylalanine, tryptophan, tyrosine and leucine. Chymotrypsin performs best in a 1:50 (w:w) enzyme to substrate ratio at basic pH and a temperature around 37°C. Chymotrypsin is also activated and stabilized by the presence of Ca^{2+} ions, therefore it is beneficial to use digestion buffers containing calcium ions (e.g. $CaCl_2$) [18]. Pepsin is an endopeptidase that is secreted in gastric chief cells as an inactive precursor called pepsinogen that becomes activated by cleavage of an N-terminal pro-segment in acidic conditions. The optimal enzymatic activity of pepsin is achieved at pH 1.5–2.5 and 37°C. Pepsin cleaves at the C-terminus of phenylalanine, leucine and rarely after histidine and lysine, unless they are adjacent to leucine or phenylalanine. Pepsin is frequently used for on-column protein digestion in hydrogen-deuterium exchange experiments (HDX), but also an application in off-line pressure assisted protein digestion has been reported [19].

Protease	Class	pH range/ion	t [°C]	Cleavage specificity	Example application	Reference
Trypsin	Serine	7–8/Ca^{2+}	37	Arg, Lys (C-term)	Primary central nervous system lymphoma	[9]
LysC	Serine	8.5	37	Lys (C-term)	Whole liver SDS lysates	[10]
LysN	Metalloproteinase	7-9/Zn^{2+}	Thermostable	Lys (N-term)	HEK 293 cells	[11]
Chymotrypsin	Serine	8/Ca^{2+}	37	Hydrophobic AAs (C-term)	Cerebrospinal fluid (CSF)	[12]
Pepsin	Aspartic	1.5-2.5	37	Preferentially Phe, Leu (C-term)	Human liver tissue	[13]
Thermolysin	Metalloproteinase	5.0–8.5 / Zn^{2+}	65–85	Ala, Met, Ile, Leu, Val, Phe (N-term)	Human liver tissue	[13]
AspN	Metalloproteinase	6.5–8.0 / Zn^{2+}	40	Asp (N-term)	Brain and liver tissue from C57BL/6 mouse	[14]
GluC	Serine	4.0, 7.8	37	Glu, Asp (C-term)	Cerebrospinal fluid (CSF), brain and liver tissue from C57BL/6 mouse	[12, 14]
ArgC	Cysteine	7.2–8.0/Ca^{2+}	37	Arg, Lys (C-term)	Cerebrospinal fluid (CSF), brain and liver tissue from C57BL/6 mouse	[12, 14]
CNBr	Chemical	—	—	Met (C-term)	Extracellular matrix of human mammary and liver tissue	[13, 15]

Table 1.
Proteases used for proteolytic digestion of protein extracts retrieved from biological material such as tissue, body fluids or cell extract. ***Table 1*** *presents the enzyme class, pH and temperature optimum, inorganic ion cofactor and specificity of protease. In addition a representative application and literature source is given.*

GluC, ArgC, LysN, AspN are also popular proteases in bottom-up proteomics as they predictably produce complementary or orthogonal peptides to trypsin with different substrate affinities. GluC is a serine protease isolated from *Staphylococcus aureus* with specificity dependent on the digestion buffer composition. For example, performing proteolysis in phosphate buffers will lead to cleavage at the C-terminus of glutamic acid and asparatic acid, but only cleavage at the C-terminus of glutamic acid will be catalysed in ammonium acetate (pH 4.0) and ammonium bicarbonate (pH 7.8) buffers [20]. GluC is known to perform optimally under pH 4.0 and pH 7.8 at 37°C while it is stable in denaturing conditions. ArgC, isolated from *Clostridium histolyticum*, is a cysteine endopeptidase cleaving at the C-terminus of arginine and sometimes at the C-terminus of lysine. Its pH optimum is 7.6 and Ca^{2+} ions also enhance its activity. ArgC digestion has recently been considered an alternative to the conventional trypsin digestion as it cleaves at the C-terminus of arginine. LysN is a metalloprotease that cleaves at the N-terminus of lysine and it is resistant to denaturation allowing digests to proceed even at temperatures higher than men tioned above. AspN is a selective metalloproteinase isolated from *Flavobacterium menigosepticum* requiring zinc atoms for its catalytical activity [21]. Its endopeptidase activity is specific to the N-terminus of aspartic acid or cysteic acid. To maintain optimal enzymatic activity it is recommended to include $ZnSO_4$ in the digestion solution buffered between pH 6.5–8.0 at a temperature of 40°C. Combining AspN with trypsin digestion increases data quality and increases protein coverage [22]. WaLP and MaLP are less known proteases cleaving at aliphatic amino acids, which makes them popular for membrane proteomic applications. Meyer et al. demonstrated that combination of data from trypsin, LysC, WaLP and MaLP digestion leads to an increase in membrane proteome coverage by 101%, compared to coverage achieved by trypsin digestion alone [23].

Broad specificity protease digestion is less common to bottom-up sample preparation, nevertheless it is used to digest rigid protein structures that resist digestion using common proteases. Proteinase K is one such serine endopeptidase isolated from fungus *Engyodontium album* that cleaves protein sequences with a broad specificity and like others discussed above requires Ca^{2+} ions for activity. Generally, it cleaves at the C-terminus of aromatic or aliphatic amino acids and is able to digest proteins in their native state or in presence of detergent such as SDS and Triton-X 100, but works best at alkaline pH 7.5–12.0 and 37°C. Most frequently, it is used for nucleic acid purification to remove proteins, but it is also suitable for some proteomic applications such as non-specific digestion of membrane proteins, protease footprinting or prion digestion. As the name implies thermolysin is a thermostable metalloproteinase isolated from *Bacillus thermoproteolyticus*. Thermolysin requires zinc and calcium ions for proteolytic activity but remains active in temperatures from 65–85°C and between pH 5.0 to 8.5. It cleaves at the N-terminus of alanine, methionine, isoleucine, leucine, valine and phenylalanine and is often used to digest proteins that resist proteolysis using conventional proteases [24]. Papain and elastase have endopeptidase activity and broad specificity that while available are rarely used in bottom-up sample preparation. Elastase is a serine endopeptidase that cleaves at the C-terminus of small hydrophobic side chains such as glycine, valine, isoleucine and leucine. While, papain is cysteine endopeptidase that cleaves at the C-terminus of arginine and lysine if it is preceded by hydrophobic amino acid, but not succeeded by valine. Subtilisin is a serine endopeptidase isolated from soil bacteria (e.g. *Bacillus licheniformis*) that is known to non-specifically cleave the peptide bond with a preference for large uncharged amino acids, although amino acids with basic side chains can be accepted in alternate binding mode [25]. Subtilisin remains active and stable under denaturing and alkaline conditions ranging from pH 8–12 and Ca^{2+} ions stabilize subtilisin structure, therefore it is essential to include $CaCl_2$

in a digestion buffer. Subtilisin's use in bottom-up proteomics is quite limited due to its wide range of specificity, nevertheless it has been reported that it could be used to reveal previously hidden areas of the proteome [26]. Cathepsins form a large group of proteases with endopeptidase activity. Their use in proteomics is not frequent but nevertheless some uses have been reported. Cathepsin L is a cysteine protease located in lysosomes, it is physiologically involved in tissue remodeling and in diseases such as cancer metastasis. Cathepsin L is catalytically active at pH 3.0–6.5 in the presence of thiol compounds [27]. Digestion using Cathepsin L has been reported in research of histone N-termini. Cathepsin C is a N-terminal dipeptidase physiologically involved in activation of serine proteases and inflammatory cells [28]. Its use in proteomic sample preparation is limited, as its cleavage is unspecific. Nevertheless, it could serve as a potent tool to generate peptides orthogonal to conventional proteases.

Thrombin is a serine protease which is proteolytically activated during the clotting process from an inactive prothrombin precursor. It is exclusively specific towards the Leu-Val-Pro-Arg-Gly-Ser motif. Therefore, it is most often used to cleave a specific linker tethered to another peptide with this sequence motif inserted into recombinant fusion protein constructs. There is a wide palette of these type of protein tag removal endopeptidases; namely Factor Xa cleaving Leu-Val-Pro-Arg-Gly-Ser motif, Enteropeptidase cleaving Asp-Asp-Asp-Asp-Lys motif, TEV Protease cleaving Glu-Asn-Leu-Tyr-Phe-Gln-Gly motif, Rhinovirus 3C Protease cleaving Leu-Glu-Val-Leu-Phe-Gln-Gly-Pro motif and several others [29]. Further details of protein tag removal proteases will not be discussed as it does not fall within scope of this chapter.

Finally, it should be noted that reproducible protein cleavage could be achieved even in non-enzymatic reactions mediated by chemical reagents. The most frequent chemical reagents to cleave peptide bond are dilute acids, such as hydrochloric acid, formic acid and acetic acid or other reagents such as cyanogen bromide (CNBr), hydroxylamine and 2-nitro-5-thiocyanobenzoate (NTCB) [30]. Exposure of proteins to dilute acids results in kinetically favored cleavage of peptide bonds at asparagine but with time others as well, while CNBr cleaves at less abundant methionine [31]. NTCB is specific towards cysteine, while hydroxyl amine reagent cleaves peptide bonds at asparagine and glycine. Generally, chemical mediated cleavage targets peptide bonds of less common amino acids producing long peptides useful in middle-down proteomics [30].

4. Technologies for analysis of limited sample amounts

Given that there is no technology to amplify proteins as may be done for nucleic acids with polymerase chain reaction, historically proteomics has faced limitations in terms of the amount of starting material required for success. Traditional proteomics approaches to sample preparation such as filter-aided sample preparation (FASP), in-gel digestion, and in-solution digestion typically require at least several micrograms of a protein sample, which can be complicated to retrieve from representative clinical samples that are by default limited in availability. Therefore, the traditional method of defining proteomes has generally produced knowledge on the underlying biology that reflect averages rendered from analysis of mixtures of cells of different types present in tissue.

As proteomics and the requisite mass spectrometry instrumentation have evolved, microscale proteomic pipelines that decrease the amount of protein required to sub-microgram levels have become available. Microscale proteomics pipelines rely on modifications of traditional proteomics pipelines frequently accompanied with cell sorting, laser capture tissue microdissection (LCM) or single cell extraction methods. Microdevices such as nano-capillary columns, microfluidic chips, miniaturised ESI introduction interfaces and miniaturised enzyme reactors are often required [32]. Introducing microscale proteomics provides a clearer picture of reality as it substantially increases sensitivity, spatial proteome resolution and leads to better understanding of how protein networks coincide on microscopic level. Despite obvious benefits, microscale proteomics still requires special instrumentation making implementation of these protocols for the moment some what difficult across laboratories worldwide.

One recent promising such technology is nanoPOTS (nanodroplet processing in one pot for trace samples) (**Figure 3A**). The nanoPOTS platform is intended for processing small cell populations in nanoliter volumes. NanoPOTS benefits from downscaling the processing volumes that in turn substantially reduces surface associated sample losses. The final step of nanoPOTS is accompanied with solid phase extraction (SPE) that concentrates, desalts and efficiently introduces a sample to nanoLC fluidics. Recently, a modification of nanoPOTS termed microPOTS was reported that is a more adoptable variant not requiring a robotic platform [33]. It has been reported that nanoPOTS could identify >3000 proteins from 10 cultured mammalian cells, while microPOTS has been reported to reproducibly identify up to 1200 and 1800 proteins from 25 HeLa cells and 50 mm square mouse liver tissue, respectively [33]. Several nanoPOTS modifications have been reported since it was

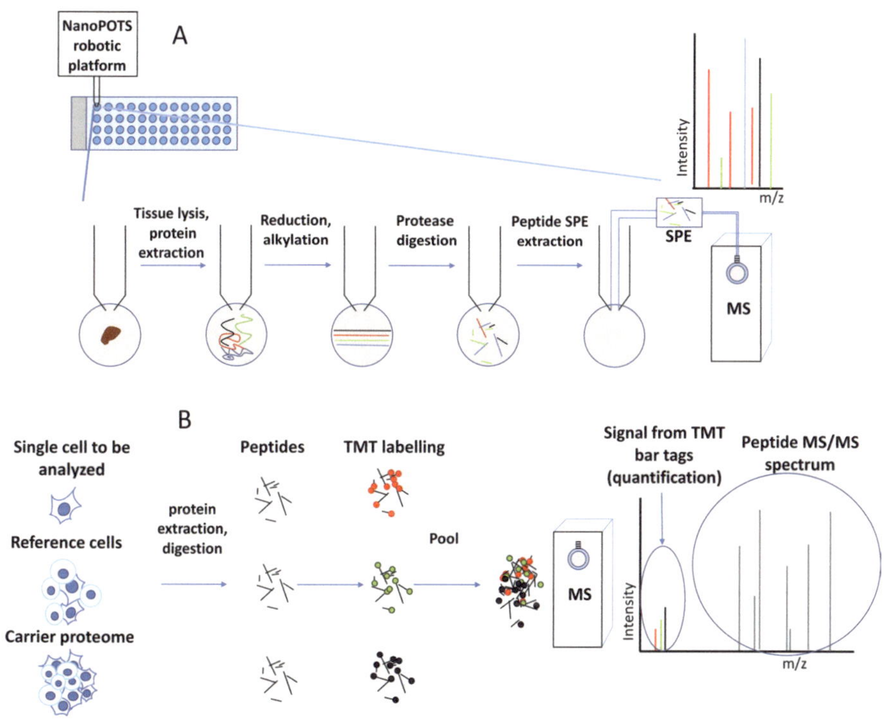

Figure 3.
Modern limited proteomic sample preparation approaches. (A) NanoPOTS; A limited proteomic sample preparation protocol that uses an automated robotic platform operating with nanoliter volumes. Sample is processed in a nano-well patterned slide. Sample preparation is based on principles of classical in-solution protein digestion. Protein digest is then transferred into SPE cartridge, where peptides are desalted and concentrated. Following, peptides are separated and analysed using mass spectrometry. (B) SCoPE-MS; a single cell proteome analysis platform. Carrier proteome is used to overcome sample losses accompanied due to peptide adsorption to surfaces. TMT labelling identifies the carrier and analysed proteomes. It could also serve for relative quantification of compared proteomes (SCoPE-MS2). Protein presence in the investigated sample and its quantity is determined based on reporter ion intensity.

introduced. For example, Zhu et al. claim that a combination of nanoPOTS with fluorescence activated cell sorting (FACS) could detect 670 protein groups from a single mammalian cell [34]. Later a combination of nanoPOTS, nanoLC separation operated at 20 nL/min and Orbitrap Eclipse and Tribrid mass spectrometer led even to a slight increase in sensitivity identifying ~1000 protein groups from a single HeLa cell [35]. Extraordinary low sample requirements predispose nanoPOTS to being useful for LC–MS/MS tissue imaging. Spatially resolved proteomic maps of a mouse blastocyst embedding into placenta have been produced using a combination of nanoPOTS and LCM. The nanoPOTS - LCM combination produced quantitative tissue images for >2000 proteins with 100-μm spatial resolution which substantially outperformed classical protein imaging mass spectrometry (IMS) [36]. The universality of nanoPOTS is well documented in several publications summarising results from pancreas, liver brain tissue thin sections as well as plant samples.

Achieving submicrogram detection limits has also been reached by introducing a carrier proteome to decrease adsorption of the proteome of interest in combination with TMT labelling (**Figure 3B**). The carrier proteome spike-in helped the method known as Single-Cell-ProtEomics-by-Mass-Spectrometry (SCoPE MS) to overcome extensive losses due to adsorption of proteins to surfaces (e.g. LC columns) while the addition of TMT labelling identifies the carrier and analysed proteomes. Moreover, TMT labels enable relative protein quantitation of multiple samples/conditions per one LC–MS run. The SCoPE MS approach has enabled detection of >1000 proteins from a single mouse embryonic stem cell [37]. Specht et al. further exploited quantitative potential of TMT labels and claimed to reproducibly quantitate >1000 proteins in a SCoPE MS experiment investigating differentiating monocytes heterogeneity [38].

Introducing on-column immobilised protease digestion (IMER) downscales sample requirements up to the sub-microgram level, especially when combined with miniaturised column diameter. Utilising various nanostructured materials such as nanoporous material, nanoparticles, nanofibers and nanotubes succeeded in IMER nanobiocatalysis as it has led to enzyme stabilisation and increasing apparent enzyme activity per unit mass of immobilisation host [39]. Several sub-microgram proteomic setups combining IMER with downstream microfluidic platforms have been reported [40–42].

The microfluidic platform termed Open tubular lab-on-column combines LysC and trypsin enzymatic digestion on 20 mm inner diameter (ID) column with on-line connected nano LC–MS/MS system. Open tubular lab-on-column benefits from very narrow capillary ID and IMER column ID that prevent excessive peptide dilution and adsorption to fluidics. The authors detected a biomarker Axin 1 in 10 ng of HCT15 colon cancer cells [40]. Huang et al. characterised 348 proteins from 25 mice blastocysts on a platform termed SNaPP coupling enzymatic digestion on 150 mm ID IMER to nanofluidics [41]. Naldi et al. coupled SCX column-based IMER proteomic reactor to nano-proteomic platform capable of protein capture, reduction, alkylation, digestion and the first dimensional SCX peptide pre-separation followed by LC–MS/MS. These authors claim that the platform performs with as low as 200 ng protein starting material [42]. Moreover, the integrated Proteome Analysis Device (iPAD) couples a 10 port valve, digestion loop and SPE trap column in a microfluidic setup that is intended for micro sample preparation prior to mass spectrometry. The authors claim that the iPAD approach is capable of identifying 813 proteins in approx. 100 Duke's type C colorectal adenocarcinoma [43].

Capillary electrophoresis (CE) is an efficient and sensitive separation technique reliably resolving proteins or peptides. Historically, it has been less robust than nanoLC but recently this has begun to change. Specifically, the introduction of CE-ESI interfaces that do not lead to an excessive peptide dilution have made

CE-MS applicable in microproteomics [44]. Several reports describe various proteomic pipelines coupling CE to MS. An ultrasensitive electrokinetically pumped nanospray ionization source coupled with CE was able to identify 283 proteins from 80 ng of MCF7 breast cancer cells. Moreover, the detection limit of spiked-in angiotensin II in bovine serum albumin digest was 2 attomole/injection [45]. Although animal proteomics does not fall within scope of this chapter it is worth mentioning that CE-MS input allowed analysis down to 50 ng of *Xenopus laevis* eggs in a single protein extract. The authors of this study used linear polyacrylamide coating and sulfonate-silica hybrid strong cation exchange monolith for SPE followed by CE-MS [46]. Combining SPE with CE in 2D manner is a promising candidate for the future development of microscale CE-MS proteomics.

5. Conclusions and future perspectives

Developments in proteomics to identify clinically relevant proteins has been widely used in scientific research. Sample preparation has been considered as one of the key steps during analysis, and as such a variety of protocols to minimize variability and to obtain best sensitivity and protein recovery from the material have been used.

Constant development of technologies that could be applied in a medical context and potentially used for screening of patient samples have been rising in recent years. Technological evolution has also had an impact to provide platforms for proteome screening of limited cell numbers, i.e. some technologies have clearly demonstrated success on the single cell level. Cellular heterogeneity at the cellular level results during tumour development that can confound analysis. Therefore, advancement of the tools for profiling of cellular subpopulations or regions of tumours has great potential to provide novel insight in mechanisms of tumour growth. Moreover, integration of developed tools with machine learning algorithms to discover and map molecules that manifest pathological development will likely lead to a better understanding of mechanisms of oncogenesis and potentially uncover therapeutic targets.

Acknowledgements

This work was supported by the International Centre for Cancer Vaccine Science, carried out within the International Research Agendas program of the Foundation for Polish Science, co-financed by the European Union under the European Regional Development Fund. The University of Victoria-Genome BC Proteomics Centre is grateful to Genome Canada and Genome British Columbia for financial support for Genomics Technology Platforms (GTP) funding for operations and technology development (264PRO).

Author details

Jakub Faktor[1], David R. Goodlett[1,2,3] and Irena Dapic[1]*

1 International Centre for Cancer Vaccine Science, University of Gdańsk, Gdańsk, Poland

2 Department of Biochemistry and Microbiology, University of Victoria, Victoria, BC, Canada

3 University of Victoria Genome BC Proteome Centre, Victoria, BC, Canada

*Address all correspondence to: irena.dapic@ug.edu.pl

References

[1] M. M. Morente et al., 'TuBaFrost 2: Standardising tissue collection and quality control procedures for a European virtual frozen tissue bank network', Eur. J. Cancer Oxf. Engl. 1990, vol. 42, no. 16, pp. 2684-2691, Nov. 2006, doi: 10.1016/j.ejca.2006.04.029.

[2] S. R. Mager et al., 'Standard operating procedure for the collection of fresh frozen tissue samples', Eur. J. Cancer Oxf. Engl. 1990, vol. 43, no. 5, pp. 828-834, Mar. 2007, doi: 10.1016/j.ejca.2007.01.002.

[3] H. Meng et al., 'Tissue triage and freezing for models of skeletal muscle disease', J. Vis. Exp. JoVE, no. 89, Jul. 2014, doi: 10.3791/51586.

[4] B. Enthaler, T. Bussmann, J. K. Pruns, C. Rapp, M. Fischer, and J.-P. Vietzke, 'Influence of various on-tissue washing procedures on the entire protein quantity and the quality of matrix-assisted laser desorption/ionization spectra', Rapid Commun. Mass Spectrom. RCM, vol. 27, no. 8, pp. 878-884, Apr. 2013, doi: 10.1002/rcm.6513.

[5] L. Giusti and A. Lucacchini, 'Proteomic studies of formalin-fixed paraffin-embedded tissues', Expert Rev. Proteomics, vol. 10, no. 2, pp. 165-177, Apr. 2013, doi: 10.1586/epr.13.3.

[6] E. I. Chen, D. Cociorva, J. L. Norris, and J. R. Yates, 'Optimization of Mass Spectrometry-Compatible Surfactants for Shotgun Proteomics', J. Proteome Res., vol. 6, no. 7, pp. 2529-2538, Jul. 2007, doi: 10.1021/pr060682a.

[7] I. Rehman and S. Botelho, 'Biochemistry, Tertiary Structure, Protein', in StatPearls, Treasure Island (FL): StatPearls Publishing, 2020.

[8] S. Suttapitugsakul, H. Xiao, J. Smeekens, and R. Wu, 'Evaluation and optimization of reduction and alkylation methods to maximize peptide identification with MS-based proteomics', Mol. Biosyst., vol. 13, no. 12, pp. 2574-2582, Nov. 2017, doi: 10.1039/c7mb00393e.

[9] Y. Zhu et al., 'High-throughput proteomic analysis of FFPE tissue samples facilitates tumor stratification', Mol. Oncol., vol. 13, no. 11, pp. 2305-2328, 2019, doi: 10.1002/1878-0261.12570.

[10] J. R. Wiśniewski, C. Wegler, and P. Artursson, 'Multiple-Enzyme-Digestion Strategy Improves Accuracy and Sensitivity of Label- and Standard-Free Absolute Quantification to a Level That Is Achievable by Analysis with Stable Isotope-Labeled Standard Spiking', J. Proteome Res., vol. 18, no. 1, pp. 217-224, 04 2019, doi: 10.1021/acs.jproteome.8b00549.

[11] N. Taouatas, S. Mohammed, and A. J. R. Heck, 'Exploring new proteome space: combining Lys-N proteolytic digestion and strong cation exchange (SCX) separation in peptide-centric MS-driven proteomics', Methods Mol. Biol. Clifton NJ, vol. 753, pp. 157-167, 2011, doi: 10.1007/978-1-61779-148-2_11.

[12] R. G. Biringer, H. Amato, M. G. Harrington, A. N. Fonteh, J. N. Riggins, and A. F. R. Hühmer, 'Enhanced sequence coverage of proteins in human cerebrospinal fluid using multiple enzymatic digestion and linear ion trap LC-MS/MS', Brief. Funct. Genomic. Proteomic., vol. 5, no. 2, pp. 144-153, Jun. 2006, doi: 10.1093/bfgp/ell026.

[13] R. Chen et al., 'Glycoproteomics analysis of human liver tissue by combination of multiple enzyme digestion and hydrazide chemistry', J. Proteome Res., vol. 8, no. 2, pp. 651-661, Feb. 2009, doi: 10.1021/pr8008012.

[14] J. R. Wiśniewski and M. Mann, 'Consecutive proteolytic digestion in

an enzyme reactor increases depth of proteomic and phosphoproteomic analysis', Anal. Chem., vol. 84, no. 6, pp. 2631-2637, Mar. 2012, doi: 10.1021/ac300006b.

[15] E. T. Goddard et al., 'Quantitative extracellular matrix proteomics to study mammary and liver tissue microenvironments', Int. J. Biochem. Cell Biol., vol. 81, no. Pt A, pp. 223-232, 2016, doi: 10.1016/j.biocel.2016.10.014.

[16] S. J. Walmsley, P. A. Rudnick, Y. Liang, Q. Dong, S. E. Stein, and A. I. Nesvizhskii, 'Comprehensive Analysis of Protein Digestion Using Six Trypsins Reveals the Origin of Trypsin As a Significant Source of Variability in Proteomics', J. Proteome Res., vol. 12, no. 12, pp. 5666-5680, Dec. 2013, doi: 10.1021/pr400611h.

[17] T. Sipost and J. R. Merkelt, 'An Effect of Calcium Ions on the Activity, Heat Stability, and Structure of Trypsin', p. 10.

[18] M. Kotormán, I. Laczkó, A. Szabó, and L. M. Simon, 'Effects of Ca2+ on catalytic activity and conformation of trypsin and α-chymotrypsin in aqueous ethanol', Biochem. Biophys. Res. Commun., vol. 304, no. 1, pp. 18-21, Apr. 2003, doi: 10.1016/S0006-291X(03)00534-5.

[19] D. López-Ferrer et al., 'Pressurized Pepsin Digestion in Proteomics: An Automatable Alternative to Trypsin for Integrated Top-Down Bottom-Up Proteomics', Mol. Cell. Proteomics, vol. 10, no. 2, p. M110.001479, Feb. 2011, doi: 10.1074/mcp.M110.001479.

[20] S. B. Sorensen, L. Sorensen, and K. Breddam, 'Fragmentation of proteins by *S. aureus* strain V8 protease', p. 3.

[21] M. A. Porzio and A. M. Pearson, 'Isolation of an extracellular neutral proteinase from Pseudomonas fragi', Biochim. Biophys. Acta BBA - Enzymol., vol. 384, no. 1, pp. 235-241, Mar. 1975, doi: 10.1016/0005-2744(75)90112-6.

[22] D. L. Swaney, C. D. Wenger, and J. J. Coon, 'Value of Using Multiple Proteases for Large-Scale Mass Spectrometry-Based Proteomics', J. Proteome Res., vol. 9, no. 3, pp. 1323-1329, Mar. 2010, doi: 10.1021/pr900863u.

[23] J. G. Meyer, S. Kim, D. A. Maltby, M. Ghassemian, N. Bandeira, and E. A. Komives, 'Expanding Proteome Coverage with Orthogonal-specificity α-Lytic Proteases', Mol Cell Proteomics, vol. 13, pp. 823-835, Mar 2014, doi: 10.1074/mcp.M113.034710.

[24] J. P. Owen, B. C. Maddison, G. C. Whitelam, and K. C. Gough, 'Use of thermolysin in the diagnosis of prion diseases', Mol. Biotechnol., vol. 35, no. 2, pp. 161-170, Feb. 2007, doi: 10.1007/BF02686111.

[25] T. P. Graycar, R. R. Bott, S. D. Power, and D. A. Estell, 'Subtilisins', in Handbook of Proteolytic Enzymes, Elsevier, 2013, pp. 3148-3155.

[26] J. Reinders, U. Lewandrowski, J. Moebius, Y. Wagner, and A. Sickmann, 'Challenges in mass spectrometry-based proteomics', PROTEOMICS, vol. 4, no. 12, pp. 3686-3703, Dec. 2004, doi: 10.1002/pmic.200400869.

[27] F. M. Dehrmann, T. H. T. Coetzer, R. N. Pike, and C. Dennison, 'Mature Cathepsin L Is Substantially Active in the Ionic Milieu of the Extracellular Medium', Arch. Biochem. Biophys., vol. 324, no. 1, pp. 93-98, Dec. 1995, doi: 10.1006/abbi.1995.9924.

[28] D. Turk, 'Structure of human dipeptidyl peptidase I (cathepsin C): exclusion domain added to an endopeptidase framework creates the machine for activation of granular

serine proteases', EMBO J., vol. 20, no. 23, pp. 6570-6582, Dec. 2001, doi: 10.1093/emboj/20.23.6570.

[29] D. S. Waugh, 'An overview of enzymatic reagents for the removal of affinity tags', Protein Expr. Purif., vol. 80, no. 2, pp. 283-293, Dec. 2011, doi: 10.1016/j.pep.2011.08.005.

[30] L. Switzar, M. Giera, and W. M. A. Niessen, 'Protein Digestion: An Overview of the Available Techniques and Recent Developments', J. Proteome Res., vol. 12, no. 3, pp. 1067-1077, Mar. 2013, doi: 10.1021/pr301201x.

[31] E. Gross, 'Nonenzymatic Cleavage of Peptide Bonds: The Methionine Residues in Bovine Pancreatic Ribonuclease', vol. 237, no. 6, p. 6, 1962.

[32] M. Safdar, J. Sproß, and J. Jänis, 'Microscale immobilized enzyme reactors in proteomics: Latest developments', J. Chromatogr. A, vol. 1324, pp. 1-10, Jan. 2014, doi: 10.1016/j.chroma.2013.11.045.

[33] K. Xu et al., 'Benchtop-compatible sample processing workflow for proteome profiling of < 100 mammalian cells', Anal. Bioanal. Chem., vol. 411, no. 19, pp. 4587-4596, Jul. 2019, doi: 10.1007/s00216-018-1493-9.

[34] Y. Zhu et al., 'Proteomic Analysis of Single Mammalian Cells Enabled by Microfluidic Nanodroplet Sample Preparation and Ultrasensitive NanoLC-MS', Angew. Chem. Int. Ed., vol. 57, no. 38, pp. 12370-12374, Sep. 2018, doi: 10.1002/anie.201802843.

[35] Y. Cong et al., 'Ultrasensitive single-cell proteomics workflow identifies >1000 protein groups per mammalian cell', Systems Biology, preprint, Jun. 2020. doi: 10.1101/2020.06.03.132449.

[36] P. D. Piehowski et al., 'Automated mass spectrometry imaging of over 2000 proteins from tissue sections at 100-μm spatial resolution', Nat. Commun., vol. 11, Jan. 2020, doi: 10.1038/s41467-019-13858-z.

[37] B. Budnik, E. Levy, G. Harmange, and N. Slavov, 'SCoPE-MS: mass spectrometry of single mammalian cells quantifies proteome heterogeneity during cell differentiation', Genome Biol., vol. 19, no. 1, p. 161, Dec. 2018, doi: 10.1186/s13059-018-1547-5.

[38] H. Specht et al., 'Single-cell mass-spectrometry quantifies the emergence of macrophage heterogeneity', bioRxiv, p. 665307, Dec. 2019, doi: 10.1101/665307.

[39] J. Kim, B. C. Kim, D. Lopez-Ferrer, K. Petritis, and R. D. Smith, 'Nanobiocatalysis for protein digestion in proteomic analysis', Proteomics, vol. 10, no. 4, pp. 687-699, Feb. 2010, doi: 10.1002/pmic.200900519.

[40] H. K. Hustoft et al., 'Open Tubular Lab-On-Column/Mass Spectrometry for Targeted Proteomics of Nanogram Sample Amounts', PLoS ONE, vol. 9, no. 9, p. e106881, Sep. 2014, doi: 10.1371/journal.pone.0106881.

[41] E. L. Huang et al., 'SNaPP: Simplified Nanoproteomics Platform for Reproducible Global Proteomic Analysis of Nanogram Protein Quantities', Endocrinology, vol. 157, no. 3, pp. 1307-1314, Mar. 2016, doi: 10.1210/en.2015-1821.

[42] M. Naldi, U. Černigoj, A. Štrancar, and M. Bartolini, 'Towards automation in protein digestion: Development of a monolithic trypsin immobilized reactor for highly efficient on-line digestion and analysis', Talanta, vol. 167, pp. 143-157, May 2017, doi: 10.1016/j.talanta.2017.02.016.

[43] Q. Chen, G. Yan, M. Gao, and X. Zhang, 'Ultrasensitive Proteome Profiling for 100 Living Cells by Direct Cell Injection, Online Digestion and

Nano-LC-MS/MS Analysis', Anal. Chem., vol. 87, no. 13, pp. 6674-6680, Jul. 2015, doi: 10.1021/acs.analchem.5b00808.

[44] R. Gahoual, E. Leize-Wagner, P. Houzé, and Y.-N. François, 'Revealing the potential of capillary electrophoresis/mass spectrometry: the tipping point', Rapid Commun. Mass Spectrom., vol. 33, no. S1, pp. 11-19, Sep. 2018, doi: 10.1002/rcm.8238.

[45] L. Sun, G. Zhu, S. Mou, Y. Zhao, M. M. Champion, and N. J. Dovichi, 'Capillary zone electrophoresis-electrospray ionization-tandem mass spectrometry for quantitative parallel reaction monitoring of peptide abundance and single-shot proteomic analysis of a human cell line', J. Chromatogr. A, vol. 1359, pp. 303-308, Sep. 2014, doi: 10.1016/j.chroma.2014.07.024.

[46] Z. Zhang, G. Zhu, E. H. Peuchen, and N. J. Dovichi, 'Preparation of linear polyacrylamide coating and strong cationic exchange hybrid monolith in a single capillary, and its application as an automated platform for bottom-up proteomics by capillary electrophoresis-mass spectrometry', Microchim. Acta, vol. 184, no. 3, pp. 921-925, Mar. 2017, doi: 10.1007/s00604-017-2084-8.

Chapter 3

Identification of Peptides and Proteins in Illegally Distributed Products by MALDI-TOF-MS

Ahmad Amini, Torgny Rundlöf, Henrik Lodén, Johan A. Carlsson, Martin Lavén, Ezra Mulugeta, Karin Björk, Torbjörn Arvidsson, Iréne Agerkvist and Anette Perolari

Abstract

An analytical strategy based on matrix-assisted laser desorption/ionization (MALDI) time-of-flight (TOF) mass spectrometry (MS) for identification of peptides and proteins in illegally distributed products is presented. The identified compounds include human growth hormone (hGH), human somatoliberin, anti-obesity drug (AOD), growth hormone releasing peptides (GHRP-2 and GHRP-6), Glycine-GHRP-2 and Glycine-GHRP-6, ipamorelin, insulin aspart and porcine, delta sleep-inducing peptide (DSIP), thymosin β4, insulin like growth factor (IGF), mechano growth factor (MGF), human chorionic gonadotropin (hCG), melanotan II, bremelanotide, dermorphin and body protecting compound (BPC 157). The identification of proteins was mainly based on peptide mass fingerprinting, i.e., *bottom up* approach, while the smaller peptides were identified through *de-novo* sequencing. In cases when a reference standard was available, complementary identification was performed by capillary electrophoresis in double-injection mode (DICE), where a suspicious product was compared with the reference standard through two consecutive injections within the same electrophoretic run.

Keywords: matrix-assisted laser desorption/ionization time-of-flight mass spectrometry, double-injection capillary electrophoresis, illegally distributed proteins and peptides

1. Introduction

A broad range of proteins and peptides, for various purposes of enhancement, such as human growth hormone (hGH), i.e., somatropin, can be obtained from the illicit market. These products are mainly marketed as lyophilized formulations in small glass containers often without labelling. The customers are exposed to a range of potential harms, besides from the active components, including bacterial and fungal or viral infections which may arise from the fact that they are administered parenterally.

Figure 1A illustrates the total number of injection vials containing white lyophilized product cake being seized by the Swedish Customs during nine years in the past, i.e., 2010–2018. A large proportion of these samples, i.e., 64%, contained

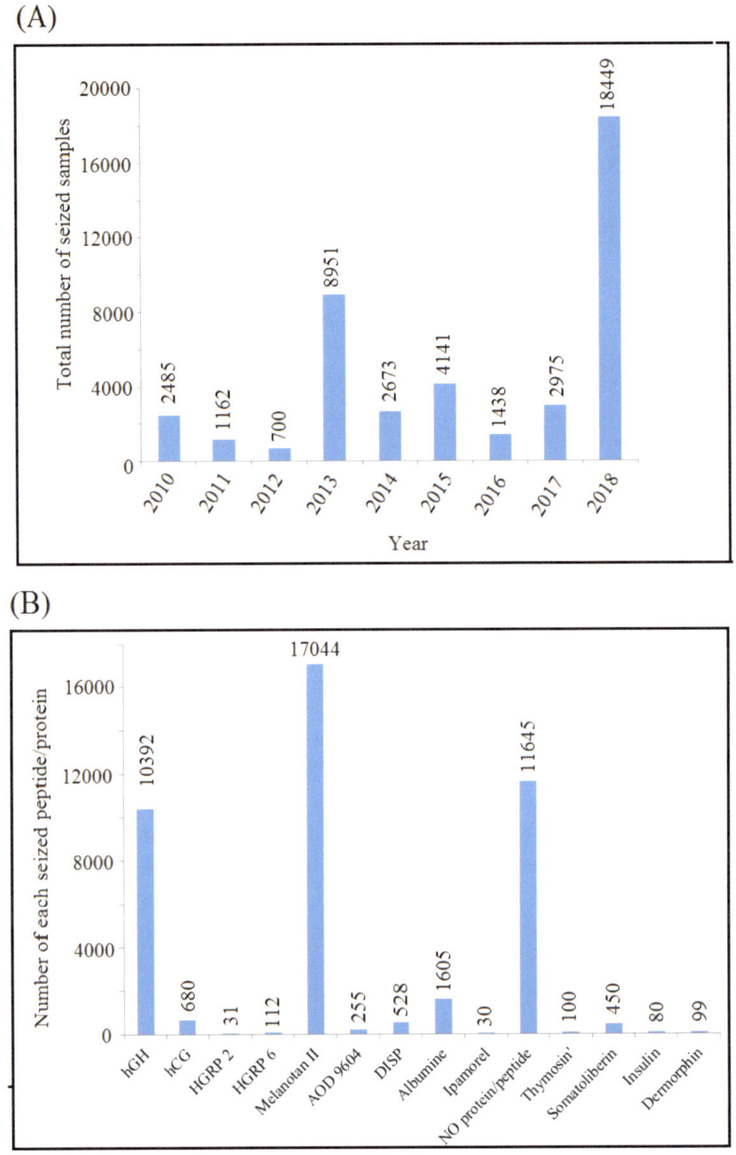

Figure 1.
Schematic illustration of the number of seized illicit products during 2010–2018 in Sweden (A), as well as the active peptides/proteins that have been identified in theses samples (B).

human growth hormone or melanotan II. About a third of the seized vials, i.e., 27%, did not contain any active peptide or protein, while the remaining 9% of the vials contained other compounds (**Figure 1B**).

The concept of a proteolytic peptide pattern, i.e., protein peptide mapping (PPM), being characteristic of a protein was first demonstrated by SDS-PAGE [1]. In 1989, peptide sequencing by automated Edman degradation had a cycle-time of nearly one hour per amino acid residue. Samples of interest often contained complex mixtures of proteins, which usually required separation by SDS-PAGE followed by electroblotting onto a polyvinylidene fluoride (PVDF) membrane [2]. However, a more rapid approach to peptide sequencing is “peptide mass fingerprinting” (PMF). By PMF, proteins are enzymatically cleaved in a predictable manner and the sizes of the generated peptide fragments are specific for different proteins.

Subsequent analysis of the obtained peptides by mass spectrometry (MS) generates mass-to-charge ratio (*m/z*) values in the mass spectrum which in turn give rise to a characteristic "peptide mass fingerprint" of the protein [3, 4]. The fingerprint serves to identify the protein by comparison with *in silico* digests, i.e., search engines attempt to match peptides from *in silico* digested proteins to those measured by the mass spectrometer [5–9]. Peptide mass fingerprinting with MS, which was first demonstrated with fast atom bombardment ionization in 1981, provides the possibility of identifying a protein at nanogram-level [5, 10–12]. Trypsin is a commonly used proteolytic enzyme for PMF, since it is relatively cheap, highly selective, and generates peptides with an average size of about 8–10 amino acids which are ideally suited for analysis by MS. It cleaves principally on the C-terminal side of arginine and lysine with the exception of Arg-Pro and Lys-Pro [2]. Limitations to protein identification by PMF include; I) The protein sequence must be present in a database for a successful protein identification. II) Proteins with extensive post-translational modifications may fail to yield good matches [13]. III) Different isoforms of a protein or alternatively spliced proteins may not be distinguished if the unique sequence regions are not observed in the peptide map. IV) Incomplete proteolytic digestion and differences in peptide ionization provide an incomplete mass fingerprint of the protein. Therefore, a complementary approach to PMF for protein identification is the use of tandem mass spectrometry (MS/MS), whereby tryptic peptide ions from the first stage of MS are dissociated along the backbone and then separated and detected in a second stage of MS to identify primary amino acid sequences [14–16]. Tandem mass spectrometry in conjunction with PMF provides even more specificity, thereby facilitating the identification [17, 18].

Since the innovation of sensitive commercial instrumentation based on MALDI-TOF MS in 1992, the technique has been widely used for protein identification due to its high sensitivity and mass accuracy, speed, extremely low material consumption, absence of multiple charge mass signals and relatively high tolerance toward additives and contaminants such as salts, matrix components and excipients [19–26]. Furthermore, MALDI is a micro-destructive analytical technique and the remaining material on the MALDI target plate can be archived for later analysis. The high sensitivity of MALDI implies that only a small aliquot of the digested protein is required for mass analysis, and the remainder can be used for alternative measurements. MALDI provides additional information regarding the primary structure of the protein by sequencing of selected tryptic peptide ions in post source decay (PSD) mode [27–34]. MALDI in-source decay (ISD) is another attractive method which generates partial sequence information of intact proteins with up to 20–50 amino acid residues [35] (**Figure 2**).

The sequence information from MALDI-PSD or MALDI-ISD analyses can be used to validate protein identification. The singly charged ions generated by MALDI-TOF-MS are a mixture of b-, y- and a-ions accompanied by ions resulting from neutral loss of ammonia or water [36–39].

PMF-based protein identification is accomplished by searching a protein sequence database using different search engines such as ProFound [40], Mascot [41], or SEQUEST [15]. A value-based scoring system has been developed that facilitates the identification without accompanying amino acid data [42, 43]. Parameters which are considered to be important for the identification include; molecular mass, protein sequence coverage and the number of matching peptides [42]. However, presence of a signature peptide, being unique for a protein, facilitates the PMF-based identifications [44]. Prior reports suggest that a minimum of four matching peptides and a sequence coverage of at least 20% is necessary for positive PMF-based protein identification [45, 46]. The other alternative strategy

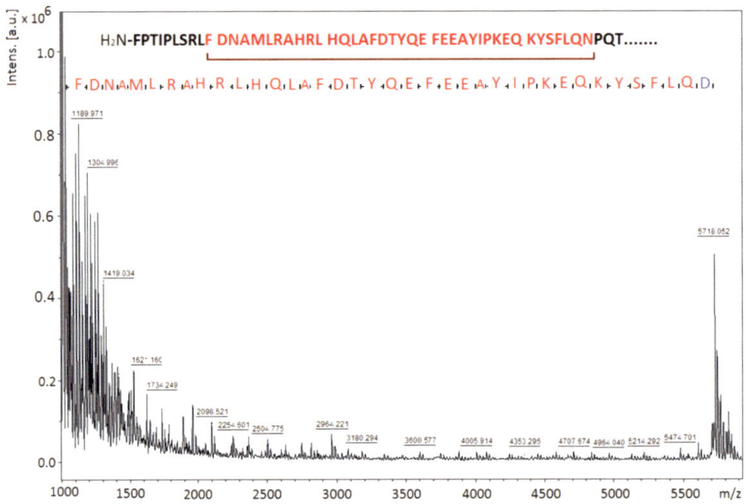

Figure 2.
MALDI in source decay analysis of a suspected illegal somatropin sample. The blue marked amino acid asp (D) is the deamidated form of Asn (N).

for protein identification is the top down approach, where intact molecule ions are subjected to gas-phase fragmentation [47].

Proteins with posttranslational modifications, such as glycosylation, present additional challenges since the masses of the modified peptides are different and thus do not contribute to the identification. In such cases, the protein can be analyzed by capillary electrophoresis (CE), in order to explore the heterogeneity of the protein followed by comparison of its electropherogram with that of the corresponding reference standard [13, 48].

2. Experimental

2.1 Sample preparation

MALDI-TOF-MS is very tolerant to salts and sample matrices, hence it is seldom necessary to desalt the sample. However, sometimes it is necessary to use a C_{18} micro-column in order to fractionate a complex sample or enhance the target analyte concentration.

The sample to be analyzed is mixed with a matrix solution (1:1, v:v), e.g. sinapinic acid (SA) or alpha-cyano-4-hydroxycinnamic acid (ACHCA). One μl of the mixture is deposited on the MALDI target plate and allowed to air-dry (i.e., the dried-droplet method) before being placed in the mass spectrometer [19, 49].

2.2 Proteolysis

The analyte to be digested is dissolved in ammonium bicarbonate (50 mM, pH 7.9). The intact sample is directly analyzed by MALDI in order to determine the molecular mass of the analyte. Then, 200 μl of the solution is digested by addition of 2–10 μl trypsin (200 μg/ml in 10 mM HCl). The reaction is carried out at room temperature or at 37°C for 30 minutes up to 24 hours, depending on peptide or protein in question. It has been found that 30 minutes digestion of somatropin at room temperature generated enough tryptic fragments for the MALDI analyses [50]. For more complex proteins, such as human chorionic gonadotropin, the required time period for proteolysis is found to be 24 hours at 37°C. Insulin porcine is digested at 37°C for 12 hours, while other peptides are digested at 37°C for 4 hours. In order to enable alkylation of the cysteine residues in a protein or peptide, it is reduced by using DTT or 2-mercaptoethanol (ME) followed by labelling of the free thiol groups with 2-iodoacetamide. The alkylation is carried out through the following procedure:

1. 2.5 µl 100 mM ME is added to 10 µl of the protein solution.

2. The protein is then incubated at 50°C for 15 minutes to reduce the S-S linkages.

3. 2.5 µl 2-iodoacetamide (100 mM) is added into the mixture to interact with free sulfide groups of the cysteine residues at +4°C for 15 to 60 minutes in darkness.

4. 2.5 µl (10 µg/mL) trypsin is added to the mixture for the digestion. The reaction is performed at room temperature or at 37°C [13, 50].

2.3 Apparatus and operating conditions

MALDI-TOF analyses are performed using either an Autoflex or an Autoflex Max (Bruker Daltonics, Bremen, Germany) reflector type time-of-flight mass spectrometer, equipped with a pulsed nitrogen laser working at 337 nm and a smartbeam II laser working at 355 nm, respectively. The Autoflex instrument is operated in the positive ion mode with delayed extraction at an accelerating voltage of 20 kV and a variable voltage reflectron. The parameter settings are optimized to analyze peptides in reflectron mode. Before analysis, the instrument is externally calibrated with Bruker Daltonics standard peptide or protein mixtures. Peptide mass peaks occurring due to autolysis of trypsin (porcine) such as 842.51 and 2211.10 Da are also used for internal calibration. Mass spectra are obtained by averaging 250 laser shots (5× 50 shots) at different positions on the sample surface. All samples being used for post source decay (PSD) analysis are analyzed in the reflectron mode. The autoflex Max instrument TOF/TOF (2 kHz MS and 200 Hz MS/MS) operates in the positive ion mode. Metastable fragmentation is induced by laser (355 nm) without the further use of collision gas. The lyophilized samples are dissolved in 300 µL ammonium bicarbonate buffer (50 mM, pH 7.5). The liquid samples are diluted with same buffer. The wells of MALDI plate are spotted with 1 µl sample/matrix solution (1:1, v:v) and allowed to air dry before being placed in the mass spectrometer. ACHCA is used for analysis of peptides. About 20 mg of ACHCA is mixed in 1 ml of ethanol: acetonitrile (ACN) (1: 1 v/v) and 0.1% trifluoroacetic acid (TFA). SA is used for protein analysis. Two different solutions of SA in water and ethanol are made as follows: 1 - Saturated solution of SA in ethanol and 0.1% TFA; 2 - Saturated solution of SA in 50% acetonitrile (ACN) and 0.1% TFA. Solution 1 is first applied on the MALDI plate on which the sample mixed with SA in 50% ACN and 0.1% TFA (1: 1) is then applied.

3. Results and discussion

Illegally distributed lyophilized or liquid products being suspected to contain pharmacologically active peptides were seized by the Swedish customs. The analyte to be identified is analyzed in both reflectron and linear modes in order to determine its molecular mass (**Figure 3**). Large peptides and proteins are then exposed to trypsin digestion in order to obtain peptide-mass map upon MALDI analysis in

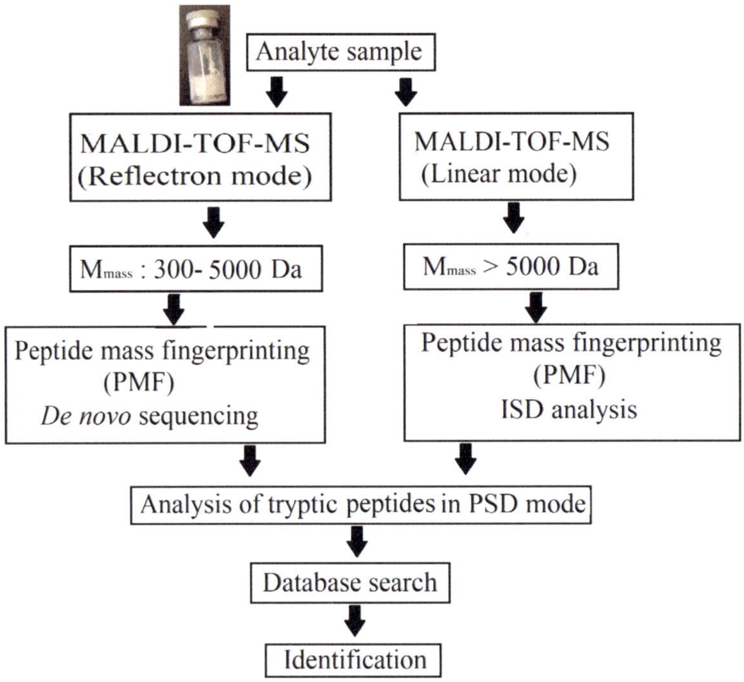

Figure 3.
The sample to be identified is analyzed in both reflectron and linear modes in order to determine the molecular mass of the analyte. Depending on the size of the molecule it will be exposed to enzymatic digestion in order to be identified through PMF. Small peptides used to be identified by de novo *sequencing in PSD mode.*

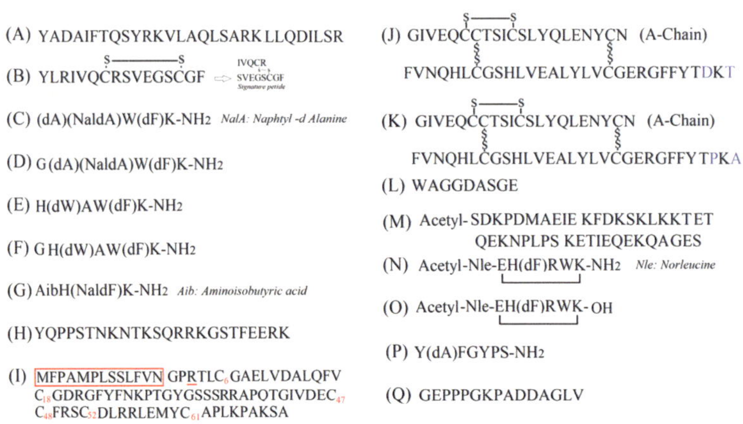

Figure 4.
The primary structure of the analyzed peptides. (A) Somatoliberin, (B) AOD, (C) GHRP-2, (D) glycine-GHRP-2, (E) GHRP-6, (F) glycine-GHRP-2, (G) Ipamorelin, (H) MGF, (I) long-R3-IGF (disulfide bridges: C_6-C_{48}; C_{47}-C_{52} and C_{18}-C_{61}; asp at position 3 is replaced by Arg), (J) insulin Aspart, (K) insulin porcine, (L) DSIP, (M) Thymosine β_4, (N) Melanotan II, (O) Bremelanotide, (P) Dermorphin and (Q) BPC 157. For molecular structures of somatropin and hCG see references [13, 50].

reflectron mode. Small peptides are, on the other hand, analyzed in reflectron mode and/or PSD mode directly. This strategy was applied to the identification of the following peptides and proteins (**Figure 4** and **Table 1**).

3.1 Identification of somatropin (hGH)

Recombinant hGH or somatropin consists of 191 amino acids with two disulfide bridges (Cys53-Cys165 and Cys182-Cys189) and promotes proteinogenesis as well as

Peptide	M_{mass}	PMF[a]	PSD[b]	ISD[c]	DICZE[d]	NMR	LC/MS
Somatropin	22115.07 22128.68[e]	X	—	X	X	X	X
Human Somatoliberin	3366.866	X	X	—	—	—	—
AOD (Anti Obesity Drug) HGH fragment 177–191	1813.850	X	X	—	—	—	—
GHRP-2	817.397	—	X	—	—	—	—
Gly-GHRP-2	874.419	—	X	—	—	—	—
GHRP-6	872.433	—	X	—	—	—	—
Gly-GHRP-6	929.455	—	X	—	—	—	—
Ipamorelin	711.385	—	X	—	—	—	—
MGF	2866.469	X	X	—	—	—	—
Long-R3-IGF	9105.385 9111.576[e]	X	X	—	—	—	—
Insulin Porcine	5772.766	X	X	—	X	X	X
Insulin Aspart	5821.611	X	X	—	X	X	X
DSIP	848.318	—	X	—	—	—	—
Thymosin-β_4	4960.474	X	X	—	—	—	—
hCG α - Chain β - Chain α + β	13,431[e] 23,114[e] 36,341[e]	X	X	—	X	—	—
Melanotan-II	1023.502	—	X	—	—	X	—
Bremelanotide	1024.510	—	X	—	—	X	—
Dermorphin	802.337	—	X	—	—	X	X
BPC-157	1418.692	—	X	—	—	X	X
Albumin bovine[f]	> 66,000[e]	X	—	—	—	—	—

[a]*Identification by peptide mass fingerprinting using enzymatic degradation as well as other modifications.*
[b]De novo *sequencing by MALDI- post source decay.*
[c]*Protein sequencing by MALDI- in source decay.*
[d]*Identification and/or impurity profiling by double injection capillary zone electrophoresis.*
[e]*Average molecular mass.*
[f]*Bovin albumin was detected in some of the samples.*

Table 1.
Illegally distributed peptides and proteins that have been analyzed by MALDI-ToF-MS and DICZE. The monoisotopic mass (M_{mass}) of the analytes and the employed analytical methodology is indicated.

fat mobilization and oxidation [51–53]. Recombinant hGH is used as a prescription drug to treat children's growth disorders and adult growth hormone deficiency. In the belief that the beneficial impact of somatropin on the growth can be extrapolated to healthy individuals, it is abused by bodybuilders and athletes [54]. However, many users are unaware of the correct dosage and how to prepare the solution for giving an injection. It has been demonstrated that supra-physiological dosages can have fatal consequences [55]. Apart from the undesired consequences following the abuse of somatropin, our investigations have shown that the illegally marketed products contained high levels of impurities such as endotoxins [50]. Endotoxins are associated with Gram-negative bacteria which can cause severe immune response and diseases in humans [56, 57]. Somatropin was identified through PMF and MALDI-ISD (see **Figure 2**) [48, 58, 59]. The availability of a compendial reference standard has made it possible to apply double injection capillary zone electrophoresis (DICZE) for both identification and impurity determination of somatropin products [50, 58, 59]. The DICZE-method provided complementary information on the native protein, providing a side by side comparison between the electrophoretic patterns of the reference standard and the analyte to be identified [50].

3.2 Identification of human somatoliberin

Human somatoliberin, growth hormone-releasing hormone (GHRH), constitutes of 44 amino acids without any post-translational modification or disulfide bridge. Somatoliberin was first isolated from two pancreatic islet cell tumors, and subsequently from normal human hypothalamus [60–62]. The MALDI results from determination of the molecular mass, PMF and amino acid sequence revealed that the Asn_8 (N), Gly_{15} (G) and Met_{27} (M) residues have, respectively, been replaced by Gln_8 (Q), Ala_{15} (A) and Leu_{27} (L) during the synthesis (see **Figures 4** and **5**). The peptide was successfully identified by PMF and *de-novo* sequencing of three of the tryptic peptides.

3.3 Identification of an anti-obesity drug (AOD)

The AOD peptide is a fragment of the C-terminus of human growth hormone (fragment 177–191) where a tyrosine is added at the N-terminus. It is a cyclic peptide consisting of 16 amino acids with a disulfide bridge between cysteine residues at positions 7 and 14 in the peptide chain [63] (**Figure 4** and **Table 1**). The fragment is the minimum length of the hGH sequence that retains the lipolytic and antilipogenic properties of hGH [63–65]. The molecular peptide masses of its tryptic peptides complied with the peptide map of hGH fragment 177–191. The existence of the disulfide bridge between C_7 and C_{14} was confirmed upon analysis of the non-reduced tryptic sample (**Figure 6**). This peptide has also been employed as a signature peptide for the identification of hGH [48, 50]. The amino acid sequences of three selected tryptic peptides were also confirmed.

3.4 Identification of growth hormone releasing peptides (GHRP)

GHRP, including GHRP-2, GHRP-6, Gly-GHRP-2, Gly-GHRP-6 and ipamorelin, as an agonist of the gut peptide ghrelin is an endogenous ligand for the growth

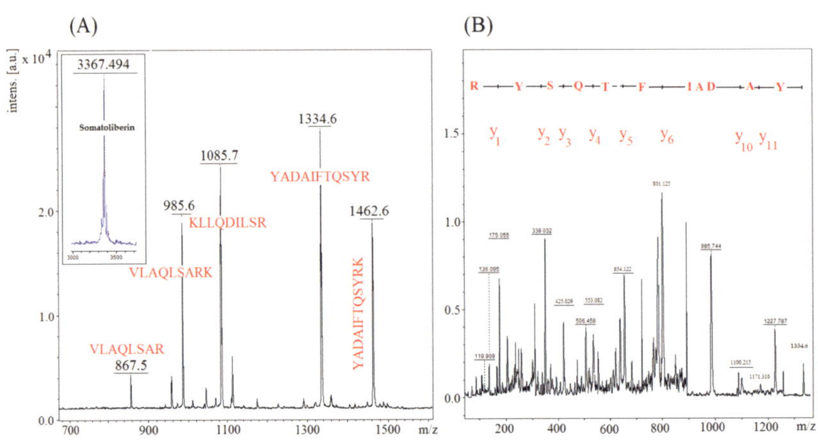

Figure 5.
MALDI-PMF (A) and MALDI-PSD (B) analysis of somatoliberin.

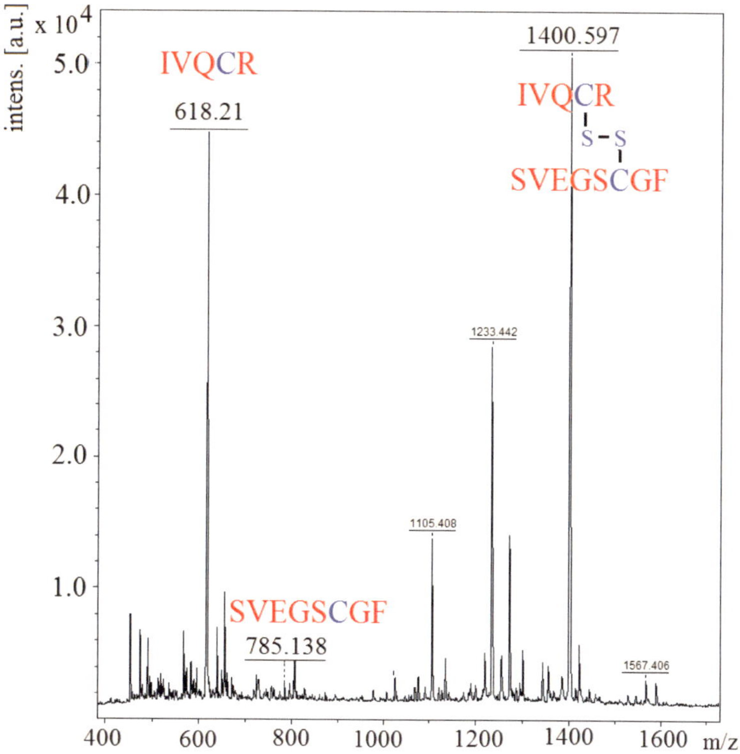

Figure 6.
MALDI-PSD analysis of AOD.

hormone secretagogue receptor [66, 67]. Ghrelin strongly stimulates food intake and GH release in humans [68–70]. These peptides were identified through *de-novo* sequencing. The amino acid sequence of GHRP-6 differs slightly from that of GHRP-2, i.e., the amino acid residues dA and Naphthyl alanine (NalA) in GHRP2 are replaced by H and dW in GHRP-6 (**Figure 4** and **Table 1**) [70].

Ipamorelin is a penta-peptide, being derived from GHRP-1 [71]. Ipamorelin like the other GHR-peptides, stimulates production of growth hormone [72]. Incorporation of aminoisobutyric acid (Aib) in the peptide chain increases the stability of the peptide (**Figure 4**) [73].

3.5 Identification of mechano growth factor (MGF) and long-R3 insulin-like growth factor (IGF-1)

MGF is a unique, spliced variant of IGF-1. MGF induces muscle cell proliferation in response to muscle stress and injury [74]. MGF and Long-R3-IGF1 were identified in several confiscated samples. Long-R3-IGF-1, an analogue of IGF-1, has 13 additional amino acids at its N-terminus (**Figure 4** and **Table 1**). IGF-1 mediates the anabolic and mitogenic activity of GH [75–77]. MGF and Long-R3-IGF1 were identified by sequence coverages of 100% and 43%, respectively (**Table 2** and **Figure 7**).

3.6 Identification of insulin porcine and insulin aspart

Insulin regulates the cellular uptake, utilization, and storage of glucose, amino acids, and fatty acids and inhibits the breakdown of glycogen, protein, and fat. Since more than one decade ago the illegal use of insulin has been noticed

Peptide fragments	Theoretical m/z $[M+H]^+$	Determined m/z $[M+H]^+$
YQPPSTNKNTKSQRRKGSTFEERK	2869.169	2869.422
Glu-C digestion of the peptide[a]:		
YQPPSTNKNTKSQRRKGSTFEE	2584.809	2584.816
Trypsin digestion of the peptide:		
GSTFEERK[b]	953.458	953.574
YQPPSTNK[b]	934.453	934.525
GSTFEER[b]	825.363	825.469
SQR[b]	390.199	390.186
NTK[b]	362.193	362.302

[a] *Glu-C cleaves at the C-terminus of either aspartic or glutamic acid residues.*
[b] *The amino acid sequence of the peptide was determined.*

Table 2.
MALDI peptide mass fingerprinting-data from analysis of mechano growth factor.

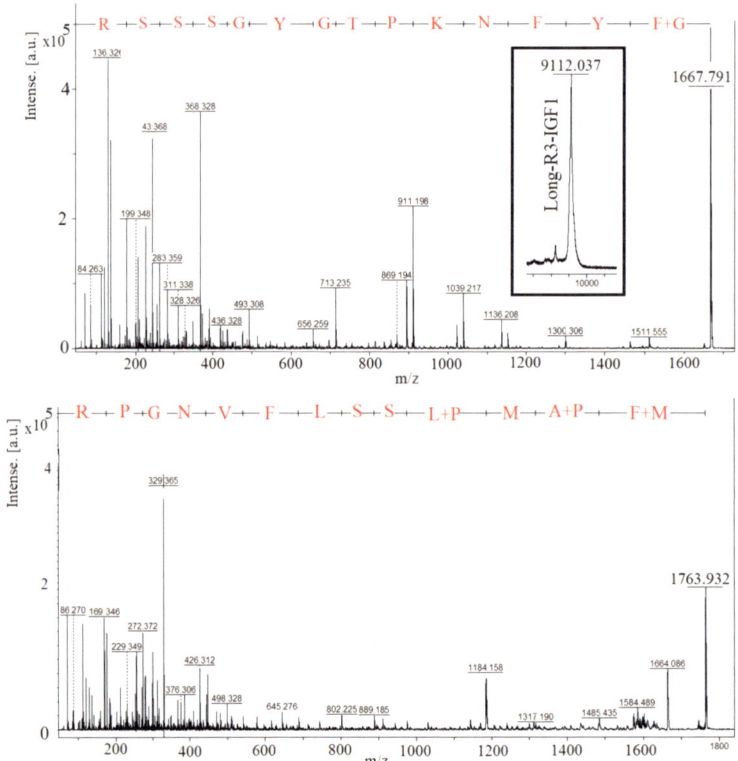

Figure 7.
MALDI analysis of intact long-R3-IGF and MALDI-PSD analysis of two tryptic peptides, i.e., m/z 1667.771 and m/z 1763.887.

[78]. However, the misuse and wrong administration of insulin could cause the, so called, dead in the bed syndrome [79]. In bodybuilding, insulin works such as testosterone or hGH to consolidate muscle tissue. Insulin also prevents breakdown of muscles and vanishes rapidly from the body, since it has a very short half-time ($t_{1/2}$) [80].

Several illegal products containing insulin porcine or aspart have been analyzed. Insulin is composed of two peptide chains, i.e., A and B, which are joined by two inter-chain disulfide bonds. The A chain also contains an intra-chain disulfide bond (**Figure 4**). The results summarized in **Table 3**, demonstrate the applied strategy for the identification of porcine and insulin aspart. The insulin molecules were reduced using a potent reducing agent, i.e., 2-mercaptoethanol (ME). MS-analysis of the reduced samples resulted in a mass spectrum consisting of several signals from both reduced A and B chains. The A and B chains generated three and four signals, respectively, corresponding to the ME-modified peptide as described in **Table 3**. It is to be noted that the amino acid residues P and A at positions 28 and 30 in the B-chain, respectively, have been replaced by D and T in insulin aspart. Therefore, these insulin molecules are distinguished upon these differences. The tryptic digestion of the B chain yielded three peptide fragments of different sizes (**Figure 8** and **Table 3**). The molecular masses of these peptides were determined accurately, and the amino acid sequence of the tryptic peptides were determined in PSD-mode.

Insulin	Theoretical m/z $[M + H]^+$	Determined m/z $[M + H]^+$
Porcine (intact)	5774.635	5774.632
Aspart (intact)	5822.612	5822.618
Peptide chains from Insulin porcine:		
$[\text{A-chain} + Na]^+$	2404.990	2404.758
$[\text{A-chain} + 1ME + Na]^{+a}$	2480.988	2480.769
$[\text{B-chain} + H]^+$	3398.682	3398.460
$[\text{B-chain} + 1ME + H]^{+a}$	3474.680	3474.486
Peptide chain from Insulin aspart:[b]		
$[\text{B-chain} + H]^+$	3446.667	3446.434
[B-chain + Na]+	3468.648	3468.487
$[\text{B-chain} + 1ME + H]^{+a}$	3522.665	3522.422
$[\text{B-chain - (GFFYTDKT)} + H]^{+c}$	2487.228	2487.030
$[\text{B-chain - (GFFYTDKT)} + 1ME + H]^{+a, c}$	2563.226	2563.302
Tryptic peptides from Insulin aspart:		
$[\text{GFFYTDK} + H]^+$	877.399	877.317
$[\text{GFFYTDKT} + H]^+$	978.457	978.446
$[\text{B-chain - (GFFYTDKT)} + H]^{+c, d}$	2487.217	2487.030
Tryptic peptides from Insulin porcine:		
$[\text{GFFYTPK} + H]^+$	859.425	859.345
$[\text{GFFYTPKA} + H]^+$	930.462	930.337
$[\text{B-chain - (GFFYTPKA)} + H]^{+c}$	2487.228	2487.234
$[\text{B-chain-(GFFYTPKA)} + 1ME + H]^{+a, c}$	2563.226	2563.129

[a]*Beta mercaptoethanol (ME) was used as reducing agent.*
[b]*The A-chains of insulin aspart and Insulin porcine are identical.*
[c]*Trypsinated B-chain.*
[d]*These peptides originate from insulin aspart, see* ***Figure 8****.*

Table 3.
MALDI-TOF-MS analysis of insulin porcine and aspart.

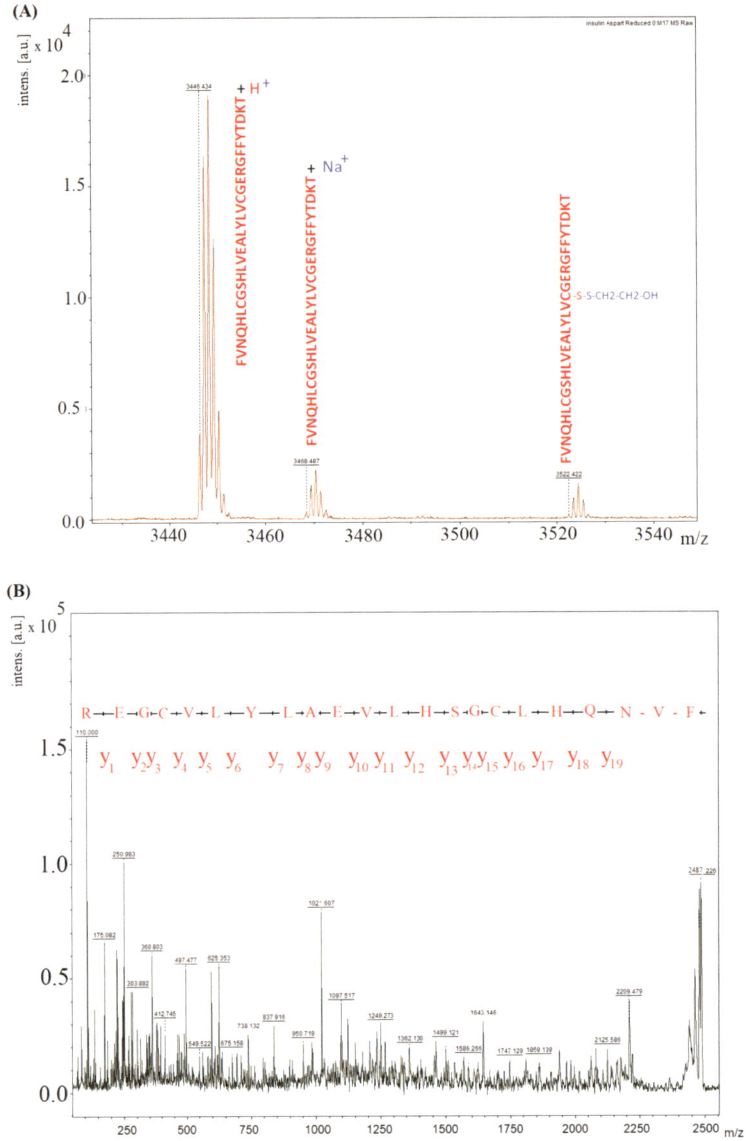

Figure 8.
*MALDI analysis of insulin aspart; analysis of reduced B-chain (A), MALDI-PSD analysis of tryptic B-chain (B), see **Table 3**.*

Double-injection capillary electrophoresis has also been applied for the identification of insulin molecules [81].

3.7 Identification of delta sleep-inducing peptide (DSIP)

The nonapeptide delta DSIP was first isolated from the cerebral venous blood of rabbits in an induced state of sleep during the mid-70s [82]. It was primarily believed to be involved in sleep regulation due to its apparent ability to induce slow-wave sleep in rabbits. However, it has been demonstrated that short-term treatment of chronic insomnia with DSIP is not likely to be of major therapeutic benefit [83]. The peptide is marketed illegally presumably for the treatment of insomnia. The peptide was directly exposed to the PSD analysis in order to confirm its molecular mass and amino acid sequence (**Figure 4** and **Table 1**).

3.8 Identification of thymosin β4

Synthetic thymosin is a peptide consisting of 43 amino acids with artificial acetylation of the N-terminus (see **Figure 4** and **Table 1**). Thymosin has the potential of playing a significant role in tissue development, maintenance, repair, pathology and other important biological activities [84]. Some important biological activities of thymosin are related to the peptide sequence $L_{17}KKTET_{22}$ [85]. Illegally distributed thymosin products are claimed to promote a variety of beneficial biological functions, such as muscle building. The peptide was identified through PMF and de-novo sequencing of the tryptic peptides (**Table 4**).

3.9 Identification of human chorionic gonadotropin (hCG)

Human chorionic gonadotropin (hCG) is a glycoprotein hormone consisting of α (92 amino acids) and β-subunits (145 amino acids) being noncovalently associated [86]. These subunits are, however, highly cross-linked internally through disulfide bridges, i.e., the α-subunit has five disulfide bridges [87], while the β-subunit has six [87, 88]. The protein is heavily glycosylated where oligosaccharides are attached to the protein backbone through asparagine and serine residues and constitute approximately 30% of the molecular mass [89]. The protein has been identified using MALDI-TOF-MS and DICZE [13, 50]. Approximately 40% of the amino acid sequence of hCG was confirmed upon PMF (**Table 5**) [13].

Peptide fragments	Theoretical m/z $[M+H]^+$	Determined m/z $[M+H]^+$
Ac-SDKPDMAEIEKFDKSKLKKTETQEKNPLPSK ETI EQE-KQAGES	4961.484	4960.987
KTETQEKNPLPSKETIEQEKQAGES	2829.401	2829.219
Ac-SDKPDMAEIEKFDKSKLK	2151.090	2151.124
Ac-SDKPDMAEIEKFDKSK	1909.911	1909.698
Ac-SDKPDMAEIEKFDK[a]	1694.784	1694.765
NPLPSKETIEQEK	1512.780	1512.768
SKLKKTETQEK	1319.743	1319.729
Ac-SDKPDMAEIEK	1304.594	1304.498
ETIEQEK	876.421	876.356
TETQEK[a]	735.342	735.356
NPLPSK[a]	655.367	655.354
FDKSK	624.325	N.D.[b]
QAGES	491.199	N.D.[b]
SKLK	475.314	N.D.[b]
FDK[a]	409.198	409.196
LKK[a]	388.282	388.286
LK[a]	260.187	260.168
SK[a]	234.135	234.151

[a]*The amino acid sequence of the peptide was determined in the PSD mode.*
[b]*N.D. = Not detected.*

Table 4.
MALDI peptide mass fingerprinting data from analysis of thymosin β_4.

Peptide fragments	Peptide position in the peptide chain	Theoretical m/z $[M+H]^+$	Determined m/z $[M+H]^+$
AYPTPLR	α-hCG; 36–42	817.446	817.482
TMLVQK	α-hCG; 46–51	719.402	719.414
STNR	α-hCG; 64–67	477.231	477.165
VTVMGGFK	α-hCG; 68–75	838.439	838.471
SK	β-hCG; 1–2	234.135	243.142
PR	β-hCG; 7–8	272.161	271.996
EPLR	β-hCG; 3–6	514.288	514.291
EPLRPR	β-hCG; 3–8	767.442	767.469
DVR	β-hCG; 61–63	389.204	389.228
FESIR	β-hCG; 64–68	651.336	651.359

Table 5.
MALDI-PMF and MALDI-PSD analysis of human chorionic gonadotropin. The identified peptides from the α and β subunits are presented in the table below.

The identification was confirmed by DICZE analysis of illegal samples together with the corresponding reference standard [13, 50].

3.10 Identification of melanotan II (MII) and bremelanotide

Melanotan, a melanocortin receptor agonist, is a cyclic-lactam bridge heptapeptide which induces melanogenesis (i.e., tanning of the skin), by activation of the MC_1 receptor, being an analogue to alpha melanocyte hormone (α-MSH) [90]. The cyclic, lactam bridged structure of MII induces increased lipophilicity (**Figure 4**) [91].

Skin-tanning products that claim to contain MII are being advertised and sold on the illicit drug market. Injection of MII can result in systemic toxicity and rhabdomyolysis [90]. Bremelanotide (formerly PT-141) is an active metabolite of MII (**Table 1**).

These peptides were identified through the *top-down* approach by MALDI in PSD mode as illustrated in **Figure 9**.

3.11 Identification of dermorphin

Dermorphin is a μ-opioid receptor-binding peptide that causes both central and peripheral effects [92] (**Figure 4** and **Table 1**). This peptide, being originally isolated from the skin of the south American tree frog *Phyllomedusa sauvagii,* is classified as one of the strongest mammalian endogenous analgesic opioids [93, 94]. Dried frog skin containing dermorphin, has been used as a therapeutic agent by the Matses tribes of the upper Amazonian basin, to treat cuts during hunting expeditions [95]. The analgetic effects of dermorphin has been demonstrated in rat, horse, dog and white sea cod [92, 94]. It has been used illegally in horse racing as a pain killing agent, allowing horses to run even if injured.

This peptide, which was detected in several samples, was identified by MALDI in the PSD mode (**Figure 10**). The molecular structure was confirmed by NMR spectroscopy.

3.12 Identification of body protecting compound 157 (BPC 157)

BPC 157 being a partial sequence of body protecting compound (BPC) (M_{mass} = 40 kDa) is a synthetic peptide, which is composed of fifteen amino acids

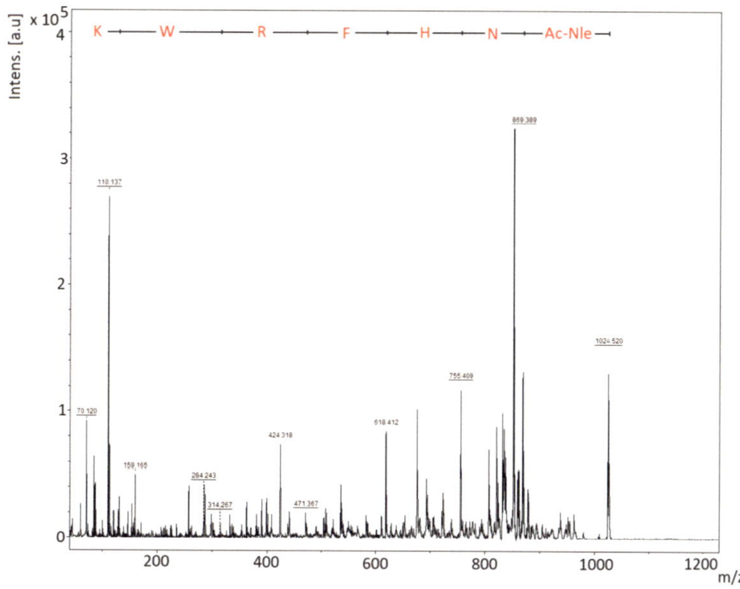

Figure 9.
MALDI-PSD analysis of melanotan II.

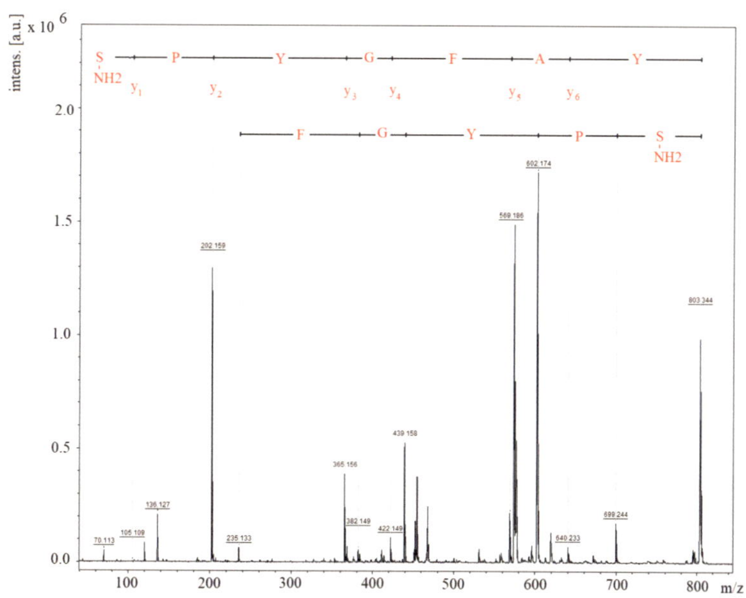

Figure 10.
MALDI-PSD analysis of dermorphin.

(**Figure 4** and **Table 1**). BPC was discovered and isolated from mouse gastric juice in response to stress stimuli in the gut mucosa [96]. BPC 157 is also known as Bepcin and PL. 14,736 or PL 10 [97]. This peptide fragment was speculated to be responsible for the BPC's physiological and protective effects [96]. However, it is unclear whether this peptide is endogenous to humans. BPC 157 is suggested to aid in tendon, ligament and muscle healing, and therefore its use as a quick injury healing in the sporting world is appealing. However, no proper clinical trials in human subjects have yet been performed to investigate the healing capability and the harmful effects of this compound [97].

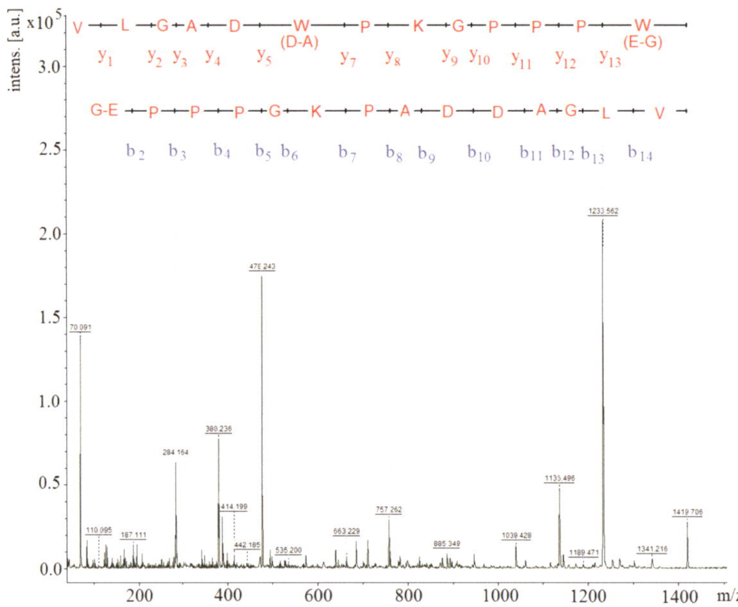

Figure 11.
MALDI-PSD analysis of BPC 157.

BPC 157 was recently identified in several confiscated vials for injection. The identification was carried out by MALDI in both PSD and reflectron modes (**Figure 11**). The amino acid sequence of the peptide was confirmed by NMR spectroscopy and LC-QTOF-MS.

4. Conclusions

The proposed methods, based on PMF by MALDI-TOF-MS as well as analysis with DICZE, provided an efficient procedure for the identification of peptides and proteins in illegally distributed samples. The use of trypsin as a proteolytic enzyme generated peptide fragments which covered 40 to 80% of the amino acid sequences of the analyzed proteins. The presence of a signature peptide in the peptide map facilitated the analyte identification considerably. MALDI-TOF-MS was also applied in the PSD mode for the amino acid sequencing of selected tryptic peptides as well as small peptides, such as ipamorelin.

The double-injection CE method provided complementary information on the native protein in the presence of a reference standard. This provided the possibility of performing a comparison between the electrophoretic patterns of the reference standard and the analyte to be identified. In addition, the double-injection based identifications were carried out by comparing the corrected migration time of the analyte and the observed migration time of the reference standard.

Abbreviations

ACHCA	α-cyano-4-hydroxycinnamic acid
ACN	Acetonitrile
Aib	Aminobutyric acid
BPC	Body protecting compound

DICZE	Double-injection capillary electrophoresis
DSIP	Delta sleep-inducing peptide
GH	Growth hormone
GHRP	Growth hormone releasing peptide
GHRH	Growth hormone releasing hormone (somatoliberin)
hCG	Human chorionic gonadotropin
hGH	Human growth hormone
IGF-1	Insulin like growth factor 1
ISD	In source decay
Nle	Norleucine
PMF	Protein mass fingerprinting
PSD	Post source decay
SA	Sinapinic acid
TFA	Trifluoroacetic acid

Author details

Ahmad Amini[1]*, Torgny Rundlöf[1], Henrik Lodén[1,2], Johan A. Carlsson[1], Martin Lavén[1], Ezra Mulugeta[1], Karin Björk[3], Torbjörn Arvidsson[1], Iréne Agerkvist[1] and Anette Perolari[1]

1 Swedish Medical Products Agency, Dag Hammarskjölds väg, Uppsala, Sweden

2 Medical Mass Spectrometry Imaging, Department of Pharmaceutical Biosciences, Uppsala University, Uppsala, Sweden

3 Swedish Customs Laboratory, Solna, Sweden

*Address all correspondence to: ahmad.amini@lakemedelsverket.se

References

[1] Cleveland DW, Fischer SG, Kirschner MW, Laemmli UK. Peptide mapping by limited proteolysis in sodium dodecyl sulfate and analysis by gel electrophoresis. Journal of Biological Chemistry. 1977; 252: 1102-1106.

[2] Matsudaira P. Sequence from picomole quantities of proteins electroblotted onto polyvinylidene difluoride membranes. Journal of Biological Chemistry. 1987;262:10035-10038.

[3] Domon B, Aebersold R. Mass Spectrometry and protein analysis. Science. 2006;312:212-217. DOI: 10.1126/science.1124619.

[4] Han X, Aslanian A, Yates III JR. Mass spectrometry for proteomics. Current Opinion in Chemical Biology. 2008;12:483-490. DOI: 10.1016/j.cbpa.2008.07.024.

[5] Yates III JR, Speicher S, Griffin PR, Hunkapiller T. Peptide mass maps: a highly informative approach to protein identification. Analytical Biochemistry. 1993, 214, 397-408. DOI: 10.1006/abio.1993.1514.

[6] Aebersold R, Goodlett DR. Mass spectrometry in proteomics. Chemical Reviews. 2001;101:269-295. DOI:org/10.1021/cr990076h.

[7] Pappin DJC, Hojrup P, Bleasby AJ. Rapid identification of proteins by peptide-mass fingerprinting. Current Biology.1993;3:327-332. DOI:10.1016/0960-9822(93)90195-t.

[8] Pandey A, Mann M. Proteomics to study genes and genomes. Nature. 2000;405:837-846. DOI:10.1038/35015709.

[9] Aebersold R, Mann M. Mass spectrometry-based proteomics. Nature. 2003;422: 198-207. DOI:10.1038/nature01511.

[10] Barber M, Bordoli RS, Sedgewick RD, Tyler AN. Fast Atom Bombardment of Solids (F.A.B.): A New ion source for mass spectrometry. Journal of the Chemical Society, Chemical Communications.1981;11:325-327. DOI: org/10.1039/C39810000325.

[11] Henzel WJ, Watanabe C. Protein identification: the origins of peptide mass fingerprinting. Journal of American Society of Mass Spectrometry. 2003;14:931-942. DOI:10.1016/S1044-0305(03)00214-9.

[12] Jensen ON, Podtelejnikov AV, Mann M. Identification of the components of simple protein mixtures by high-accuracy peptide mass mapping and database searching. Analytical Chemistry. 1997;69:4741-4750. DOI: org/10.1021/ac970896z.

[13] Durgaryan A, Rundlöf T, Lavén M, Amini A. Identification of human chorionic gonadotropin hormone in illegally distributed products by MALDI-TOF mass spectrometry and double-injection capillary zone electrophoresis. Analytical Methods. 2016;8:4188-4196. DOI:org/10.1039/C6AY01078D.

[14] Shevchenko A, Sunyaev S, Loboda A, Shevchenko A, Bork P, Ens W, Standing K G. Charting the proteomes of organisms with unsequenced genomes by MALDI-Quadrupole Time-of-Flight Mass Spectrometry and BLAST homology searching. Analytical Chemistry. 2001;72:1917-1923. DOI:org/10.1021/ac0013709.

[15] Eng JK, McCormack AL, Yates JR. An approach to correlate tandem mass spectral data of peptides with amino acid sequences in a protein database.

Journal of American Society of Mass Spectrometry. 1994;5:976-989. DOI. org/10.1021/jasms.8b00502.

[16] Zubarev RA, Kelleher NL, McLafferty FW. Electron Capture Dissociation of Multiply Charged Protein Cations. A nonergodic process. Journal of American Chemical Society. 1998;120:3265-3266. DOI:org/10.1021/ja973478k.

[17] Shevchenko A, Jensen ON, Podtelejnikov AV, Sagliocco F, Wilm M, Vorm O, Mortensen P, Shevchenko A, Boucherie H, Mann M. Linking genome and proteome by mass spectrometry: Large-scale identification of yeast proteins from two dimensional gels. Proceedings of the National Academy of Sciences. 1996;93:14440-14445. DOI:org/10.1073/pnas.93.25.14440.

[18] Damodaran S, Wood TD, Nagarajan P, Rabin RA. Evaluating peptide mass fingerprinting-based protein identification. Genomics proteomics bioinformatics. 2007;5:152-157. DOI:org/10.1016/S1672-0229(08)60002-9.

[19] Karas M, Hillenkamp F. Laser desorption ionization of proteins with molecular masses exceeding 10,000 daltons. Analytical Chemistry. 1988;60:2299-3201. DOI:org/10.1021/ac00171a028.

[20] Pan C, Xu S, Zhou H, Fu Y, Ye M, Zou H. Recent developments in methods and technology for analysis of biological samples by MALDI-TOF-MS. Analytical and Bioanalytical Chemistry Research. 2007;387:193-204. DOI: 10.1007/s00216-006-0905-4.

[21] Tanaka M, Waki H, Ido Y, Akita S, Yoshida T. Protein and polymer analysis up to *m/z* 100,000 by laser ionization time-of-flight mass spectrometry. Rapid Communication in Mass Spectrometry. 1988;2:151-153. DOI:org/10.1002/rcm.1290020802.

[22] Hrabák J, Chudáčková E, Walková R. Matrix-assisted laser desorption ionization–time of flight (MALDI-TOF) mass spectrometry for detection of antibiotic resistance mechanisms: from Research to Routine Diagnosis. Clinical Microbiology Reviews. 2013;26:103-114. DOI: 10.1128/CMR.00058-12.

[23] Henzel WJ, Billeci TM, Stults JT, Wong SC, Grimley C, Watanabe C. Identifying proteins from two-dimensional gels by molecular mass searching of peptide fragments in protein sequence databases. Proceedings of the National Academy of Sciences USA.1993;90:5011-5015. DOI:10.1073/pnas.90.11.5011.

[24] James P, Quadroni M, Carafoli E, Gonnet G. Protein identification by mass profile fingerprinting. Biochemical and Biophysical Research Communications. 1993;195:58-64. DOI: 10.1006/bbrc.1993.2009.

[25] Amini A, Dormady SJ, Riggs L, Regnier FE, The impact of buffers and surfactants from micellar electrokinetic chromatography on matrix-assisted laser desorption ionization (MALDI) mass spectrometry of peptides. Effect of buffer type and concentration on mass determination by MALDI-time-of-flight mass spectrometry. Journal of Chromatography A. 2000;894:345-355. DOI: 10.1016/s0021-9673(00)00798-6.

[26] Padliya ND, Wood TD. Improved peptide mass fingerprinting matches via optimized sample preparation in MALDI mass spectrometry. Analytica Chimica Acta. 2008;627:162-168. DOI:org/10.1016/j.aca.2008.05.059.

[27] Nilsson E, Amini A, Wretlind B, Larsson A. Pseudomonas aeruginosa infections are prevented in cystic fibrosis patients by avian antibodies binding *Pseudomonas aeruginosa* flagellin. Journal of Chromatography B.

2007;856:75-80. DOI: 10.1016/j.jchromb.2007.05.029.

[28] Spengler B, Kirsch D, Kaufmann R, Jaeger E. Peptide sequencing by matrix-assisted laser-desorption mass spectrometry. Rapid Communications in Mass Spectrometry. 1992;16:105-108. DOI: 10.1002/rcm.1290060207.

[29] Doucette A, Craft D, Li L. Protein concentration and enzyme digestion on microbeads for MALDI-TOF peptide mass mapping of proteins from dilute solutions. Analytical Chemistry. 2000;27:3355-3362. DOI: 10.1021/ac000176j.

[30] Kaufmann R, Spengler B, and Luetzenkirchen F. Mass Spectrometric Sequencing of Linear Peptides by Product-ion Analysis in a reflectron time-of-flight mass spectrometer using matrix-assisted laser desorption ionization. Rapid Communications in Mass Spectrometry. 1993;7:902-910. DOI: 10.1002/rcm.1290071010.

[31] Spengler B. Post-source decay analysis in matrix-assisted laser desorption/ionization mass spectrometry of biomolecules. Journal of Mass Spectrometry. 1997, 32, 1019-1036. DOI: 10.1002/(SICI)1096-9888(199711)32:10<1019::AID-JMS595>3.0.CO;2-G.

[32] Purcell AW, Gorman JJ. The use of post source decay in matrix assisted laser desorption-ionisation mass spectrometry to delineate T cell determinants. Journal of Immunological Methods. 2001;249:17-31. DOI: 10.1016/s0022-1759(00)00361-6.

[33] Talbo GH, Suckau D, Malkoski M, Reynolds EC. MALDI-PSD-MS analysis of the phosphorylation sites of caseinomacropeptide. Peptides. 2001;22:1093-1098. DOI: 10.1016/s0196-9781(01)00426-0.

[34] Hoteling AJ, Owens KG. Improved PSD and CID on a MALDI TOFMS. Journal of the American Society of Mass Spectrometry. 2004;15:523-535. DOI: 10.1016/j.jasms.2003.12.004.

[35] Demeure K, Quinton L, Gabelica V, De Pauw E. Rational selection of the optimum MALDI matrix for top-down proteomics by in-source decay. Analytical Chemistry. 2007;*79*:8678-8685. DOI: org/10.1021/ac070849z.

[36] Wattenberg A, Organ AJ, Schneider K, Tyldesley R, Bordoli R, Bateman RH. Sequence dependent fragmentation of peptides generated by MALDI quadrupole time-of-flight (MALDI Q-TOF) mass spectrometry and its implications for protein identification. Journal of the American Society for Mass Spectrometry. 2002;13:772-783. DOI: 10.1016/S1044-0305(02)00414-2.

[37] McCormack AL, Somogyi A, Dongré AR, Wysocki VH. Fragmentation of protonated peptides: surface-induced dissociation in conjunction with a quantum mechanical approach. Analytical Chemistry. 1993;65:2859-2872. DOI: 10.1021/ac00068a024.

[38] Dongré AR, Somogyi A, Wysocki VH. Surface-induced Dissociation: An effective tool to probe structure, energetics and fragmentation mechanisms of protonated peptides. Journal of Mass Spectrometry. *1996*;**31:**339-350. DOI: 10.1002/(SICI)1096-9888(199604)31:4<339::AID-JMS322>3.0.CO;2-L.

[39] Tsaprailis G, Nair H, Somogyi A, Wysocki VH, Zhong W, Futrell JH, Summerfield SG, Gaskell SJ. Influence of secondary structure on the fragmentation of protonated peptides. Journal of the American Chemical Society. 1999;**121**:5142-5154. DOI: 10.1021/ja982980h.

[40] Zhang WZ, Chait BT. ProFound: an expert system for protein identification

using mass spectrometric peptide mapping information. Analytical Chemistry. 2000;72:2482-2489. DOI: 10.1021/ac991363o.

[41] Perkins DN, Pappin DJC, Creasy JM, Cottrell JS. Probability-based protein identification by searching sequence databases using mass spectrometry data. Electrophoresis. 1999;20:3551-3567. DOI: 10.1002/(SICI)1522-2683(19991201)20:18<3551::AID-ELPS3551>3.0.CO;2-2.

[42] Damodaran S, Wood TD, Nagarajan P, Rabin RA. Evaluating peptide mass fingerprinting-based protein identification. Genomics, Proteomics & Bioinformatics. 2007;5:152-157. DOI: 10.1016/S1672-0229(08)60002-9.

[43] Holcapek M, Jirásko R, Lísa M. Recent developments in liquid chromatography-mass spectrometry and related techniques. Journal of Chromatography A. 2012;1259:3-15. DOI: 10.1016/j.chroma.2012.08.072.

[44] Ji J, Chakraborty A, Geng M, Zhang X, Amini A, Bina M, Regnier FE. Strategy for qualitative and quantitative analysis in proteomics based on signature peptides. Journal of Chromatography B. 2000;745:197-210. DOI: 10.1016/S0378-4347(00)00192-4.

[45] Biron DG, Brun C, Lefevre T, Lebarbenchon C, Loxdale HD, Chevenet F, Brizard JF, Thomas F. The pitfalls of proteomics experiments without the correct use of bioinformatics tools. Proteomics. 2006;6:5577-5596. DOI: 10.1002/pmic.200600223.

[46] Barrett J, Brophy PM, Hamilton JV. Analysing proteomic data. International Journal for Parasitology. 2005;35:543-553. DOI: 10.1016/j.ijpara.2005.01.013.

[47] Sze SK, Ge Y, Oh H, McLafferty FW. Top-down mass spectrometry of a 29-kDa protein for characterization of any posttranslational modification to within one residue. Proceedings of the National Academy of Sciences. 2002;99:1774-1779. DOI: 10.1073/pnas.251691898.

[48] Amini A. Separation of somatropin charge variants by multiple-injection CZE with Polybrene/chondroitin sulfate A double-coated capillaries. Journal of Separation Science. 2013;36:2686-2690. DOI: 10.1002/jssc.201300327.

[49] Avinash AP, Cheng-Kang C, Chen-Hao W, Wen-Ping P. Forced dried droplet method for MALDI sample preparation. Analytica Chimica Acta. 2018;1031:128-133. DOI:org/10.1016/j.aca.2018.05.056.

[50] Amini A, Rundlöf T, Rönnquist K, Tydén M, Turunen T, Korhola P and Perolari A. A protocol for the quality assessment of illegally distributed human growth hormones with respect to identity, purity, endotoxin level and microorganism content. Analytical Methods. 2015;7: 8857-8864. DOI: org/10.1039/C5AY01390A.

[51] Hober FF and Haymond MW. Human growth hormone prevents the protein catabolic side effects of prednisone in humans. Journal of Clinical Investigation. 1990;86: 265-272. Doi: 10.1172/JCI114694.

[52] Dupin P, Galinou F and Bayol A. Analysis of recombinant human growth hormone and its related impurities by capillary electrophoresis. Journal of Chromatography A. 1995;707: 396-400. Doi.org/10.1016/0021-9673(95)00324-G.

[53] Somatropin for injection. European Pharmacopoeia (Ph. Eur.). 0951 (01/2008).

[54] Crist DM, Peake GT, Egan PA and Waters DL. Body composition response to exogenous GH during training in highly conditioned adults Journal of

applied Physiology. 1988;65: 579-584. DOI: 10.1152/jappl.1988.65.2.579.

[55] Takala J, Ruokonen E, Webster NR, Nielsen MS, Zandstra DF, Vundelinckx G and Hinds CJ. Increased mortality associated with growth hormone treatment in critically ill adults. The New England Journal of Medicine. 1999;341: 785-792. DOI: 10.1056/NEJM199909093411102.

[56] McCuskey R, Urbaschek R and Urbaschek B. The microcirculation during endotoxemia. Cardiovascular Research. 1996;32: 752-763. Doi.org/10.1016/S0008-6363(96)00113-7.

[57] Carcillo JA and Cunnion RE. Septic shock. Critical Care Clinics. 1997;13:553-574. DOI: 10.1016/s0749-0704(05)70328-6.

[58] Amini A. Rapid identification of somatropin by peptide-mass fingerprinting, using MALDI-TOF mass spectrometry. Pharmeuropa Bio & Scientific Notes, 2009;1: 11-6.

[59] Webster J and Oxley D. Peptide mass fingerprinting: protein identification using MALDI-TOF mass spectrometry. Methods in Molecular Biology. 2005;310: 227-240. DOI: 10.1007/978-1-59259-948-6_16.

[60] Guillemin R, Brazeau P, Bohlen PE, Esch R, Ling N and Wehrenberg W. Growth hormone-releasing factor from a human pancreatic tumor that caused acromegaly. Science. 1982;218: 585-587. DOI: 10.1126/science.6812220.

[61] Rivier J, Spiess JL, Thorner MO and Vale W. Characterization of a growth hormone-releasing factor from a human pancreatic islet tumour. Nature. 1982;300: 276-278. DOI: 10.1038/300276a0.

[62] Ling N, Esch R, Böhlen P, Brazeau P, Wehrenberg W and Guillemin R. Isolation, primary structure, and synthesis of human hypothalamic somatocrinin: growth hormone-releasing factor. Proceedings of the National Academy of Sciences of the USA. 1984;81: 4302-4306. DOI: 10.1073/pnas.81.14.4302.

[63] Ogru E, Wilson JC, Heffernan M, Jiang WJ, Chalmers DK, Libinaki R and Ng FM. The conformational and biological analysis of a cyclic anti-obesity peptide from the C-terminal domain of human growth hormone. Journal of Peptide Science. 2000;56: 388-397. DOI: 10.1034/j.1399-3011.2000.00771.x.

[64] Wijaya E and Ng FM. Effect of an antilipogenic fragment of human growth hormone on glucose transport in rat adipocytes. Biochemistry and molecular biology international. 1993;31: 543-552.

[65] Ho KK, O'Sullivan AJ and Hoffman DM. Metabolic Actions of Growth Hormone in Man. Endocrine Journal. 1996;43: S57-S63. DOI: 10.1507/endocrj.43.suppl_s5.

[66] Bowers CY, Momany FA, Reynolds GA and Hong A. On the in vitro and in vivo activity of a new synthetic hexapeptide that acts on the pituitary to specifically release growth hormone. Endocrinology. 1984;114: 1537-1545. DOI: 10.1210/endo-114-5-1537.

[67] Kojima M, Hosoda H, Date Y, Nakazato M, Matsuo H and Kangawa K. Ghrelin is a growth-hormone-releasing acylated peptide from stomach. Nature. 1999;402: 656-660. DOI: 10.1038/45230.

[68] Takaya K, Ariyasu H, Kanamoto N, Iwakura H, Yoshimoto A, Harada M, Mori K, Komatsu Y, Usui T, Shimatsu A, Ogawa Y, Hosoda K, Akamizu T, Kojima M, Kangawa K. and Nakao K. Ghrelin strongly stimulates growth hormone release in humans. The Journal of Clinical Endocrinology and Metabolism.

2000;85: 4908-2911. DOI:10.1210/jcem.85.12.7167.

[69] Wren A M, Small CJ, Ward HL, Murphy KG, Dakin CL, Taheri S, Kennedy AR, Roberts GH, Morgan, DGA, Ghateli MA, Bloom SR. The Novel Hypothalamic Peptide Ghrelin Stimulates Food Intake and Growth Hormone Secretion. Endocrinology. 2000;141: 4325-4328. DOI: org/10.1210/endo.141.11.7873.

[70] Popovic V, Damjanovic S, Micic D, Djurovic M, Dieguez C and Casanueva FF. Blocked growth hormone-releasing peptide (GHRP-6)-induced GH secretion and absence of the synergic action of GHRP-6 plus GH-releasing hormone in patients with hypothalamopituitary disconnection: evidence that GHRP-6 main action is exerted at the hypothalamic level. The Journal of Clinical Endocrinology and Metabolism. 1995;80: 942-947. DOI: 10.1210/jcem.80.3.7883854.

[71] Semenistaya E, Zvereva I, Thomas A, Thevis M, Krotov G and Rodchenkov G. Determination of growth hormone releasing peptides metabolites in human urine after nasal administration of GHRP-1, GHRP-2, GHRP-6, Hexarelin, and Ipamorelin. Drug Testing and Analysis. 2015;7: 919-925. DOI: 10.1002/dta.1787.

[72] Isidro *M and* Cordido F. Growth Hormone Secretagogues. Combinatorial Chemistry & High Throughput Screening, 2006;9:175-180. DOI:10.2174/138620706776055458.

[73] Banerjee A, Pramanik A, Bhattacharjya S and Balaram P. Omega amino acids in peptide design: incorporation into helices. Biopolymers. 1996;39: 769-777. DOI: 10.1002/(SICI)1097-0282(199612)39:6%3C769::AID-BIP4%3E3.0.CO;2-T.

[74] Kandalla PK, Goldspink G, Butler-Browne G, Mouly V. Mechano Growth Factor E peptide (MGF-E), derived from an isoform of IGF-1, activates human muscle progenitor cells and induces an increase in their fusion potential at different ages. Mechanisms of Aging and Development. 2011;132:154-162. DOI:10.1016/j.mad.2011.02.007.

[75] Laron Z. Biodrugs. Somatomedin-1 (recombinant insulin-like growth factor-1): clinical pharmacology and potential treatment of endocrine and metabolic disorders. 1999;11:55-70. DOI: 10.2165/00063030-199911010-00006.

[76] Green H, Morikawa M, Nixon T. A dual effector theory of growth-hormone action. Differentiation, 1985;29:195-198. DOI:org/10.1111/j.1432-0436.1985.tb00316.x.

[77] Dubaquié Y, Mortensen DL, Intintoli A, Hogue DA, Nakamura G, Rancatore P, Lester P, Sadick MD, Filvaroff E, Fielder PJ, Lowman HB. Selective Insulin-Like Growth Factor I Variants: Engineering, Biodistributions, and Clearance. Endocrinology. 2001;142: 165-173. DOI: org/10.1210/endo.142.1.7864.

[78] Sonksen PH. Insulin, growth hormone and sport. Journal of Endocrinology. 2001;170:13-25. DOI: 10-1677/joe.0.1700013.

[79] Weston PJ. The dead in bed syndrome revisited: a review of the evidence. Diabetes Management. 2012;2:233-241.

[80] Duckworth WC, Bennett RG, Hamel FG. Insulin Degradation: Progress and Potential. Endocrine Reviews. 1998;19:608-624. DOI:org/10.1210/edrv.19.5.0349.

[81] Amini A. Double-injection capillary electrophoresis for the identification of analytes. Electrophoresis. 2014;35:2915-2921. DOI:org/10.1002/elps.201400229.

[82] Schoenenberger GA, Maier PF, Tobler HJ, Monnier M. A naturally occurring delta-EEG-enhancing nonapeptide in rabbits. Pflügers Archiv -European Journal of Physics. 1977;369;99-109.

[83] Bes F, Hofman W, Schuur J, Van Boxtel C. Effects of Delta Sleep-Inducing Peptide on Sleep of Chronic Insomniac Patients. Neuropsychobiology. 1992;26:193-197. DOI:10.1159/000118919.

[84] Sosne G, QiuP, Goldstein AL, Wheater M. The biological activities of thymosin β4 defined by active sites in short peptide sequences. The FASEB Journal. 2010;24: 2144-2151. DOI:org/10.1096/fj.09-142307.

[85] Mannherz HG, Hannappel E. The beta-thymosins: intracellular and extracellular activities of a versatile actin binding protein family. Cell Motility and the Cytoskeleton. 2009;66:839-851. DOI: 10.1002/cm.20371.

[86] Stenman UH, Tiitinen A, Alfthan H, Valmu L. The classification, functions and clinical use of different isoforms of HCG. Human Reproduction Update 2006;12:769-784. DOI:org/10.1093/humupd/dml029.

[87] Mise T, Bahl OP. Assignment of disulfide bonds in the alpha subunit of human chorionic gonadotropin. Journal of biological Chemistry.1980;255:8516-8522.

[88] Mise T, Bahl OP. Assignment of disulfide bonds in b sub-unite of human chorionic gonadotropin. Journal of Biological Chemistry. 1981;256:6587-6592.

[89] Elliott MM, Kardana A, Lustbader JW, Cole LA. Carbohydrate and peptide structure of the alpha- and beta-subunits of human chorionic gonadotropin from normal and aberrant pregnancy and choriocarcinoma. Endocrine. 1997;7:15-32. DOI: 10.1007/BF02778058.

[90] Nelson ME, Bryant SM, Aks SE. Melanotan II injection resulting in systemic toxicity and rhabdomyolysis. Clinical Toxicology. 2012, **50**,1169-1173. DOI: 10.3109/15563650.2012.740637.

[91] Hadley ME. Discovery that a melanocortin regulates sexual functions in male and female humans. Peptides, 2005;26:1687-1689. DOI:org/10.1016/j.peptides.2005.01.023.

[92] Robinson MA, Guan F, McDonnell S, Uboh CE, Soma LR. Pharmacokinetics and pharmacodynamics of dermorphin in the horse. Journal of Veterinary Pharmacology and Therapeutics, 2014;38:321-329. DOI:org/10.1111/jvp.12179.

[93] Broccardo M, Usenko AB, Uranova MG, Guzevatykh LS, Kamensky AA, Andreeva LA, Alfeeva LY, Myasoedov NF, Giannini E, Improta G, Emel'yanova TG. In vitro and in vivo opioid activity of [DPro^{6}]dermorphin, a new dermorphin analogue. Peptides, 2003;24:419-428. DOI:org/10.1016/S0196-9781(03)00057-3.

[94] Chervova LS, Lapshin DN, Kamenskii AA. Pain Sensitivity of Trout and Analgesia Induced by Intranasal Administration of Dermorphin. Doklady Biological Sciences.1994;338:424-425.

[95] Negri L, Melchiorri P, Lattanzi R. Pharmacology of amphibian opiate peptides. Peptides, 2000;21:1639-1647. DOI: 10.1016/s0196-9781(00)00295-3.

[96] Sikirić, P Petek M, Ručman R, Seiwerth S, Grabarević Z, Rotkvić, I, Turković B, Jagić V, Mildner B, Duvnjak M, Lang N, Danilović Z, Cviko A, Kolega M, Sallmani A, Djačić S, Bura M, Brkić T, Banić M, Dodig M, Corić V,

Šimičević V, Veljaca M, Erceg, D, Ježek D, Simunić-Banek L, Skroza N, Bulić K, Buljat G, Hanževački M, Orihovać V, Lučinger D, Culig J, Separović J, Marović, A, Miše S, Suchanek E, Matoz W, Perović D, Gjurašin M, Mikulandra S, Derniković K, Cuk V, Karakas I. A new gastric juice peptide, BPC. An overview of the stomach-stress-organoprotection hypothesis and beneficial effects of BPC. The Journal of Physiology. 1993;87: 313-327. DOI: org/10.1016/0928-4257(93)90038-U.

[97] Cox HD, Millera GD, Eichnera D. Detection and in vitro metabolism of the confiscated peptides BPC 157 and MGF R23H. Drug Testing and Analysis. 2017;9:1490-1498. DOI: 10.1002/dta.2152.

Section 3

Mass Spectrometry in Clinical Laboratory

Chapter 4

Protoporphyrin IX Analysis from Blood and Serum in the Context of Neurosurgery of Glioblastoma

Anna Walke, Eric Suero Molina, Walter Stummer and Simone König

Abstract

Protoporphyrin IX (PPIX) is formed from δ-aminolevulinic acid (ALA) during heme biosynthesis. Due to its cyclic tetrapyrrole core structure, it absorbs in the visible region of the electromagnetic spectrum and is thus colored. Both ALA and PPIX have become of great interest to neurosurgery, because in high-grade glioma, ALA diffuses into the tumor and is converted to PPIX. Fluorescence-guided resection (FGR) takes advantage of both the enrichment of PPIX in the tumor and its fluorescent properties, which enable visualization of tumor tissue. ALA-mediated FGR thus maximizes the extent of resection with better prognosis for patients. Tumor cells are able to produce porphyrins naturally or after administration of ALA, which is also reflected in elevated plasma fluorescence of cancer patients. PPIX might thus serve as a biomarker for monitoring of the tumor burden. A liquid chromatography-mass spectrometry (LC-MS)-based method is presented to quantify PPIX in blood and serum in the context of current fluorescence-based diagnostics. The method is able to distinguish between zinc PPIX, a component of red blood cells of importance in the detection of lead poisoning and iron deficiency anemia, and metal-free PPIX. In a proof-of-principle study, it was used to follow a time course of a glioblastoma patient undergoing surgery and confirmed elevated PPIX blood levels before ALA administration. During surgery, these blood levels increased about four-fold. The here developed 10 min reversed-phase LC-target MS method now allows patient screening with high specificity and throughput.

Keywords: protoporphyrin IX, zinc protoporphyrin IX, iron deficiency, lead poisoning, porphyria, biomarker, cancer, ion trap mass spectrometry, brain tumor, glioblastoma, blood

1. Introduction

Cyclic tetrapyrroles are biologically important molecules as they form the core structure of prosthetic groups such as porphyrins (e.g. heme), natural pigments like chlorophyll, and of vitamers (cobalamins, e.g. vitamin B12). Due to their large conjugated systems, they absorb in the visible region of the electromagnetic spectrum and are thus colored [1, 2].

Porphyrins occur naturally as metal complexes with the red pigment in blood cells, heme, being the best-known example. Hemes (**Figure 1**), the cofactors of

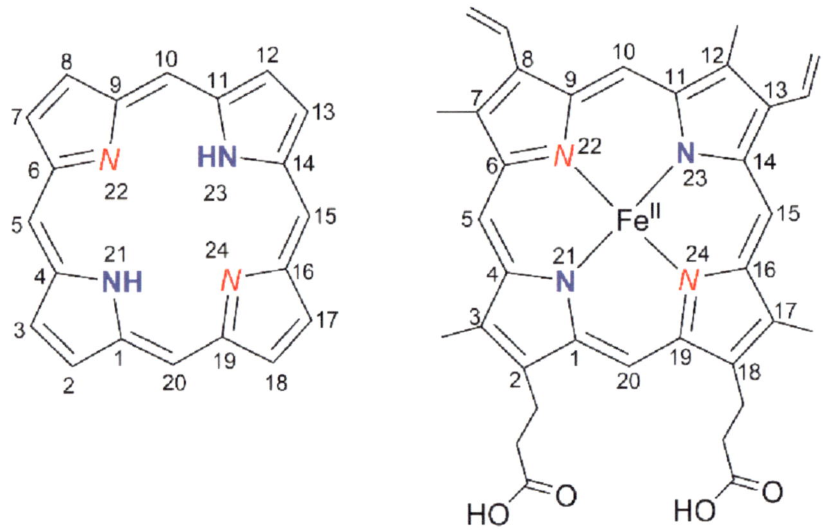

Figure 1.
Porphyrin structures. Porphin (left), the simplest representative, and heme b (right), the prosthetic group of, e.g., hemoglobin, myoglobin, catalase, and cytochromeP450. Other members of the heme group (e.g. heme c in cytochrome C) differ slightly in the side chains. Porphyrins chelate divalent ions such as iron in heme. Due to their delocalized system of π-electrons they fluoresce after excitation. Different nitrogen forms in the pyrrole ring are labelled (red/italic, blue/bold).

hemoproteins (e.g. hemoglobin, myoglobin) [3, 4], are complexed to iron and occur ubiquitously. They are critical to life, because hemoproteins are involved in the transport of diatomic gases (respiration), chemical catalysis and electron transfer [5].

During heme synthesis from glycine and succinyl-CoA (**Figure 2**) a number of intermediates including δ-aminolevulinic acid (ALA) are produced until, ultimately, protoporphyrin IX (PPIX) is converted to heme by insertion of a divalent iron (Fe (II), catalyzed by ferrochelatase) [1].

Both ALA and PPIX have become of great interest to neurosurgery, because in gliomas, ALA diffuses into the tumor and induces PPIX-synthesis [6]. A surgical method has been developed taking advantage of both the enrichment of PPIX in the tumor

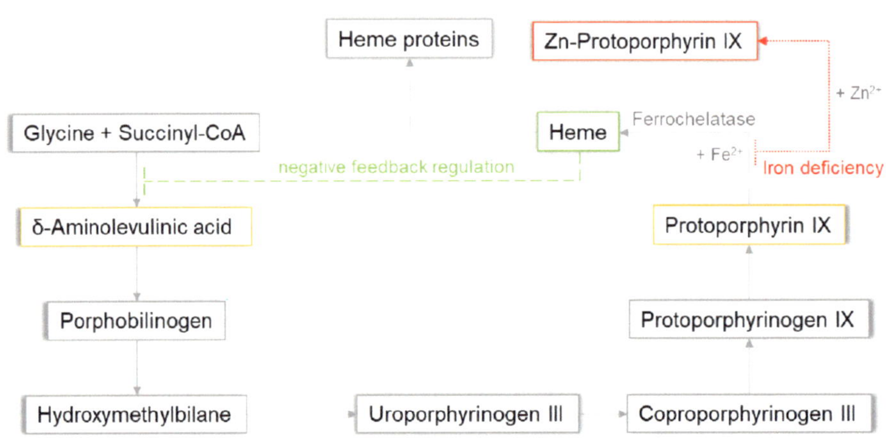

Figure 2.
Heme synthesis from succinyl-CoA to PPIX and ultimately heme. Synthesis of ALA is the rate-limiting step and under negative feedback control of heme (green/ dashed). A deficiency of iron supply limits heme synthesis and leads to PPIX accumulation. Zinc can then substitute for iron and ferrochelatase catalyzes the formation of zinc PPIX (red/dotted) [1].

and its fluorescent properties, which enable visualization of the tumor area (fluorescence-guided resection (FGR)) [7]. Alternating between normal white light illumination and violet-blue excitation light, the vital tumor tissue can be identified by its red/pink color during surgery; the distinction of infiltration zone and healthy brain tissue is improved. ALA-FGR thus maximizes the extent of tumor resection with improved prognosis for patients. In this context, the question arose whether or not PPIX blood levels could be used to monitor tumor regrowth, because the spectral characteristics of blood from cancer patients differ from those of control subjects [8–11].

The use of a simple analytic procedure would be cost-effective at lower strain for the patient. We have thus used liquid chromatography coupled to mass spectrometry (LC-MS) to quantify PPIX in blood and serum and describe it in the context of current practice in PPIX diagnostics below.

2. Protoporphyrin IX detection

2.1 Erythrocyte protoporphyrin analysis

Erythrocyte protoporphyrin (EP) served as a diagnostic marker for lead poisoning and environmental lead pollution as well as for iron deficiency anemia at the end of the 20th century [12]. From 1972 to 1991, it was officially recommended as the primary screening test for childhood lead poisoning by the Center for Disease Control and Prevention in the United States [13–15]. For the clinical diagnosis of porphyrias [16], rare disorders resulting from enzyme variability in heme biosynthesis, the porphyrin pattern is determined in blood, urine and faeces based on fluorescence techniques.

Taking advantage of the strong absorption of porphyrins in the Soret band (380– 430 nm) and their fluorescence, spectrophotometric and -fluorometric methods have been preferred for EP determination so far. The free erythrocyte porphyrin (FEP) test [17, 18], was, however, based on liquid-liquid extraction (LLE) at acidic pH, which dissociated zinc protoporphyrin (ZnPPIX) to metal-free PPIX during the extraction process. Thus, a sum parameter with different - and unknown - contributions of free PPIX and ZnPPIX was measured leading to false conclusions. The ratio of ZnPPIX to metal-free PPIX in erythrocytes varies, because in lead poisoning and iron deficiency anemia, ZnPPIX is accumulated in the blood, whereas in protoporphyria, the metal-free PPIX is elevated [12, 19, 20].

Hence, neutral ZnPPIX-specific LLE methods were developed, but they suffered from poor extraction efficiency [20, 21]. The widely applied ethyl acetate-acetic acid LLE method had three problems [17, 18]: First, the low extraction efficiency of PPIX from whole blood in comparison to the extraction from pre-diluted blood, which provided better precision of analysis; second, impurities in ethyl acetate influencing fluorescence and requiring pre-tests of reagent batches; third, the instability of PPIX standards prepared with deionised water. Thus, EP analysis required great attention to detail, because method modification, sample contamination or aging of standards and reagents had a great impact on analysis [22]. As a result, inter-laboratory comparison of EP results was generally poor while intra-laboratory precision was good [23].

2.2 Spectrophotometry and –fluorometry

In 1977, a hematofluorometer (HF) was designed for the detection of ZnPPIX in a drop of whole blood without sample pretreatment, which allowed immediate, simple and inexpensive detection [24]. Spectrophotometric and -fluorometric analysis became conventional analytical practice, but the inter-laboratory agreement for EP levels was still poor and standardization has not been achieved [25]. PPIX levels measured with the FEP-test and HF-values of ZnPPIX did not match [13, 23]. Problems arose as a result of the limited PPIX stability towards light exposure, its tendency to form molecular aggregates in aqueous solution and buffer/solvent-dependent variation in absorbance [13]. Consequently, different values have been published for the molar absorptivity of PPIX, leading to discrepancies in the calculation of PPIX and ZnPPIX concentrations from measured absorbance using Beer's Law or fluorescence emission intensity [13]. In addition, interferences such as bilirubin, increased hemoglobin, riboflavin, quinine as well as several drugs including doxorubicin or amoxycillin disturbed HF measurement; results improved with extended washing of erythrocytes [13, 26].

The application of dual-wavelength excitation in the HF technique [24] allowed the recording of background fluorescence, which then could be removed from the analyte spectra [27]. In 2019, it was shown that even the non-invasive measurement of erythrocyte ZnPPIX in children and women after childbirth was possible on the wet vermillion of the lower lip using this principle [28, 29].

The fluorescence method is rapid and easy to use, but it is limited by high background fluorescence. Complex algorithms for the processing of the obtained spectra are required. As fluorescence is measured as a sum parameter of all contributing analytes, results can easily be overinterpreted, especially when measuring at the lower limit of detection. This has to be taken into account when free PPIX is the target instead of ZnPPIX or the sum parameter of both. ZnPPIX is the predominant species of EP in circulating erythrocytes while free PPIX is only present in trace amounts [20, 30, 31]; the specific quantification of metal-free PPIX next to ZnPPIX is thus challenging.

2.3 Liquid chromatography

The absorption in the Soret band is broad and the resolution of absorption spectra is low resulting in spectral overlap. Chromatographic separation of porphyrins is thus recommended [25]. It isolates the analyte from the sample matrix and concentrates it. Conclusively, the application of high-performance liquid chromatography (HPLC) coupled with fluorescence detection (FLD, UV-visible detection was not sensitive enough) [30, 32–37] and MS became highly popular. Initially performed thin layer chromatography assays were gradually replaced in the 1980s [36].

HPLC-FLD was demonstrated in the differentiation of porphyrin species in porphyria [33, 37] or lead poisoning [36]. It is now the technique of choice for porphyrin analysis in routine and research laboratories [28, 36]. Methods are time-consuming with typical elution times above 20 min [34, 35]. Moreover, the analysis of lipophilic PPIX remains challenging, because of its comparatively low recovery. Porphyrins respond quite differently in HPLC and attempts to measure them all in a single run have been abandoned. Porphyrins also have different excretion patterns in the body so that the majority of laboratories determines free PPIX and ZnPPIX in whole blood or plasma, and other porphyrins (uroporphyrin I and III, coproporphyrin I and III, **Figure 3**) in urine, using individual HPLC-FLD methods optimized for each purpose [28, 33–35, 37].

Advanced column media and higher pressure LC instrumentation (ultra high-performance (UHP) LC) further improved resolution and analysis time at lower solvent consumption although conventional HPLC was still widely applied. UHPLC from biological matrices like blood or tissue is not yet a routine technique in porphyrin analysis, because advanced sample preparation is required to avoid column contamination and clogging. Still, the use of UHPLC in conjunction with MS has great potential regarding sensitivity and speed of analysis [38].

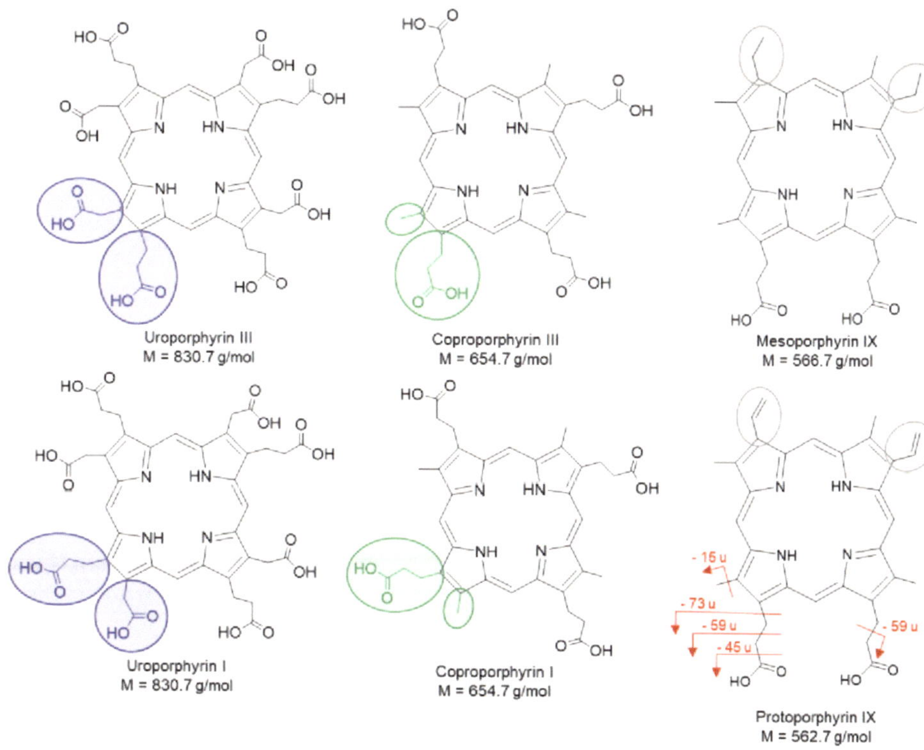

Figure 3.
Structures of type I and III isomers of uro- and coproporphyrin in comparison to PPIX and mesoporphyrin (MPIX). Differences of type I and III isomers are highlighted in blue (uroporphyrin) and green (coproporphyrin). Differences of MPIX and PPIX structures at positions 8 and 13 are shown in grey. Fragmentation of PPIX in MS analysis is given in red (in the ion trap, cleavage of ethylic side chains at positions 8 and 13 was not observed, for MS spectra see ***Figure 4****).*

2.4 Mass spectrometry

MS detection has a number of advantages over FLD [26, 27, 39]. Both high mass accuracy and the possibility of breaking the analyte molecules in the gas phase and measuring their fragments (MS/MS) contribute to unequivocal species identification, because complex samples can often not be fully chromatographically resolved. HPLC-MS was applied for the determination of porphyrin profiles in biological matrices [11, 39–43] using predominantly electrospray ionization (ESI) interfaces. Additionally, there were efforts to explore atmospheric pressure chemical ionization [42] for the purpose.

In urine, uroporphyrin, coproporphyrin and other porphyrins were measured [41, 42]. As urine is a matrix with comparatively low complexity compared to blood and tissue, minimal sample pretreatment (acidification) was sufficient allowing high-throughput analysis. Providing excellent sensitivity and specificity by operating in multiple reaction monitoring mode, the presented method was superior to FLD [41]. Another approach in urine used porphyrin esterification followed by extraction and single ion monitoring MS [42].

Regarding porphyrin detection in more complex matrices, Sullivan and co-workers [39] quantified twelve porphyrins including PPIX in liver extracts based on their mass, because most porphyrins co-eluted. MS/MS was not carried out. The method suffered from problems with peak-splitting and matrix suppression [39]. Furthermore, highly acidic extraction was performed so that this approach was not suitable for the individual determination of metal-free PPIX and ZnPPIX.

For the investigation of porphyrins in blood, predominantly plasma or red blood cells were used [11, 40, 43]. In plasma, MS was applied for the quantification of coproporphyrin isomers for monitoring of drug interactions [43], the elucidation of fluorescing compounds after detection of elevated fluorescence [11], and the qualitative analysis of porphyrin patterns facilitating the differential diagnosis of human porphyrias [40]. Despite all these efforts, no short and sensitive HPLC-MS/MS method for specific PPIX quantification from whole blood or serum was yet available although great data have been shown for less complex cell culture extracts [44, 45].

3. Protoporphyrin IX: a potential biomarker for cancer screening

More recently, besides the analysis of porphyrin metabolites [39] and profiles for toxicological and pharmacological applications [40–42], PPIX has been investigated as tumor marker for bladder, colorectal and kidney cancer [10, 11, 32]. Tumor cells are able to produce porphyrins naturally or after administration of ALA, which is also reflected in elevated plasma fluorescence of cancer patients. The spectral characteristics of blood from normal control subjects differ significantly from those of cancer patients in renal cell carcinoma, prostate cancer and colorectal adenocarcinoma [8–11].

PPIX analysis is, however, not straightforward in a clinical setting. Factors such as unrelated diseases and medication may influence the measured porphyrin concentration [8]. Lualdi and co-workers [11], e.g., confirmed their findings of enhanced plasma fluorescence in colorectal adenocarcinoma patients by HPLC coupled to high-resolution MS and detected mainly PPIX and coproporphyrin I. Ota et al. [32] applied HPLC-FLD for the determination of PPIX in plasma of bladder cancer patients after ALA administration. The patients showed significantly higher plasma PPIX concentrations compared to healthy adults. It was extrapolated that the accumulation of PPIX in cancer cells is common to almost all types of cancer [8–10] and that the specific measurement of PPIX is advantageous for cancer screening [32].

A further application of PPIX is above-mentioned photodynamic diagnosis, where PPIX is applied as an intraoperative marker especially for brain tumors. Using ALA-induced PPIX-fluorescence in tissue during surgery of high-grade glioma, the resection is more complete and the patients have a higher 6-month progression-free survival compared to those without FGR [7]. Unfortunately, due to the infiltrative growth of these tumors, complete tumor resection is still impossible and tumors can recur. Clinically, diagnosis of high-grade glioma and glioblastoma multiforme (GBM) as well as their recurrence requires multidisciplinary strategies such as contrast enhancement magnetic resonance imaging, computer tomography and biopsy [6, 7, 46, 47]. Therefore, a sensitive and cost-effective method for tumor monitoring is highly desirable supporting early diagnosis and treatment of GBM as well as better prognosis for patients. So far, the survival prognosis for GBM patients is one of the lowest in modern day oncology [47]. As PPIX is an approved marker for GBM tissue in ALA-FGR, here, the hypothesis was tested if it could also be a blood biomarker for GBM screening and diagnosis.

4. PPIX quantification

For the detection of PPIX in whole blood or serum, we developed an HPLC-MS method using an HP1100 HPLC (Agilent, Waldbronn, Germany) coupled to an Esquire 3000 ion trap mass spectrometer (Bruker Corp., Bremen, Germany).

Mesoporphyrin (MPIX) (**Figure 3**) was chosen as internal standard (IS), because it provided high structural similarity to PPIX and isotope labeled standards for PPIX were not available. Distinction of PPIX from ZnPPIX was possible during sample preparation. The method described below allowed the quantification of metal-free PPIX in whole blood, the determination of endogenous PPIX in serum and the measurement of endogenous ZnPPIX in whole blood (200 μl, respectively).

4.1 Sample preparation

PPIX LLE extraction from serum and whole blood was achieved with only water and acetonitrile (ACN). Hemolysis with water was crucial for good recovery as observed by others working with pre-dilution [22]. It was followed by protein precipitation with ACN; concomitant porphyrins were extracted into the supernatant [48], which was further purified using anionic-exchange solid phase extraction (SPE) cartridges. The extracts of whole blood and serum had a pH 8–9 so that PPIX, ZnPPIX and MPIX had deprotonated propionic acid side chains and were negatively charged. No conversion of ZnPPIX into metal-free PPIX was observed before loading the extracts onto the SPE cartridge. All three porphyrins were retained on the cartridge presenting quaternary ammonium groups. MPIX and PPIX were eluted using ACN containing 2% formic acid (FA), ZnPPIX with increased FA content (20%). No elution or hydrolysis of ZnPPIX was detected at 2% FA; only metal-free PPIX was seen in the first eluate. The higher percentage of FA in the second step caused the acidic release of the Zn^{2+} ion. ZnPPIX was thus detected as metal-free PPIX.

4.2 Gas phase fragmentation

As already demonstrated in the literature for other porphyrins [40], PPIX ionizes as singly charged $[M+H]^+$ species (*m*/z 563.3) in ESI-MS. MS/MS fragmentation preferentially occurs on the side chains (**Figure 3**). The spectra in **Figure 4** illustrate the stepwise fragmentation of PPIX from preselected precursor ions in subsequent MS/MS and MS/MS/MS experiments in the ion trap. The abundant loss of an ethanoic acid substituent (-CH_2COOH; 563-59 u) results in a fragment ion at *m*/*z* 504.3 that is used as precursor for MS^3 fragmentation which then generates further side chain losses (-CH_3, 15/30 u; -COOH, 45 u; -CH_2COOH, 59 u; -CH_2CH_2COOH, 73 u).

For selected compounds, it was discussed that even-electron ions generated in the ESI source can produce radical cations with odd-electrons by hemolytic cleavage. The most common process in radical fragmentation is the elimination of a methyl group as proposed for flavonoids, antraquinones and terpenoids [49]. It was shown that the radical elimination of the methyl group is a low energy process in flavonoids. The loss of 59 u by radical cleavage was also already described for FePPIX in previous studies on metalloporphyrins and other compounds with extended π-electron systems [50–52].

The fragmentation of MPIX is similar to PPIX. MPIX has two saturated ethyl side chains at positions 8 and 13 (**Figure 3**) so that precursor and fragment ions differ by four mass units in comparison to PPIX.

ZnPPIX was measured in MS/MS mode as it showed lower ionization efficiency than MPIX and PPIX. The time-segmented method switched from MS/MS mode for ZnPPIX detection to MS^3 mode for MPIX and PPIX measurement. This approach proved advantageous in comparison to continuous MS/MS and MS^3 switching for porphyrins and significantly increased sensitivity. In comparison to the ion traces of matrix-free porphyrin standard solution, the background signal was about five times higher in whole blood extracts, but that did not hamper detection with the specific MS^3 method.

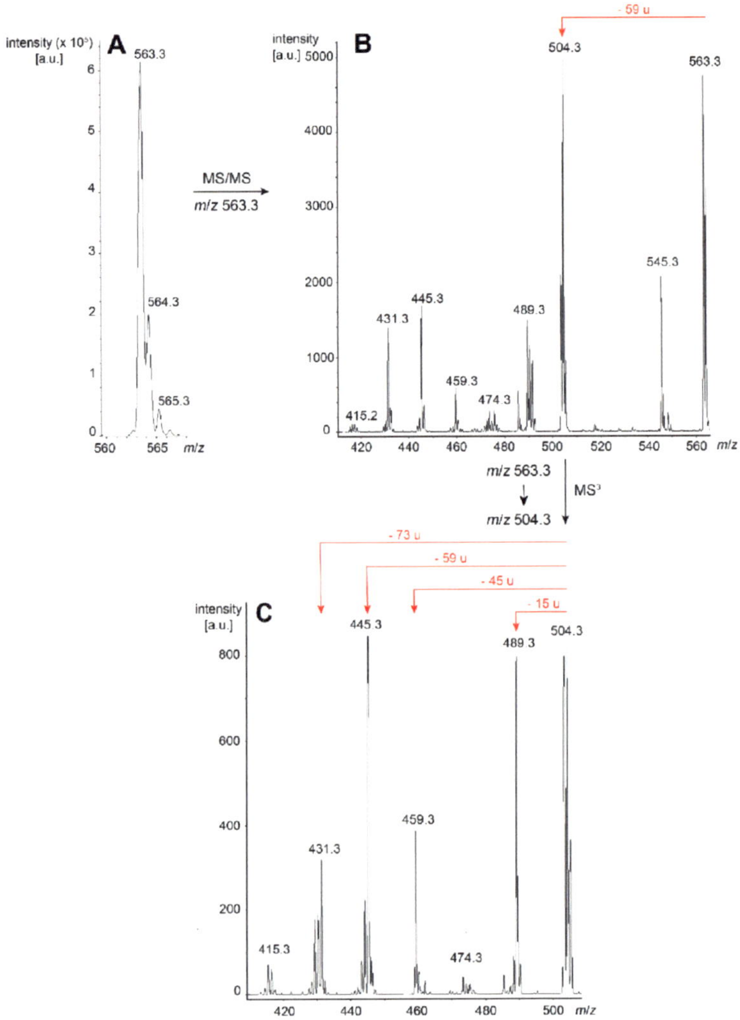

Figure 4.
Fragmentation of PPIX in the ion trap. A: $[M+H]^+$ ion of PPIX at m/z 563.3 in full scan mode. B: MS/MS spectrum of precursor ion at m/z 563.3. The loss of the ethanoic acid side chain (59 u) is dominant. C: MS^3 spectrum for two-step fragmentation (m/z 563.3 -> m/z 504.3). The major ions result from side chain cleavage (red arrows, for structure see ***Figure*** *3).*

4.3 Chromatography

HPLC separation of PPIX was not straightforward, because of its high lipophilic nature. Problems included low resolution on capillary C_8 LC and high carry-over on endcapped C_{18} phase. The low flow rate of capillary LC (5 µl/min) was also disadvantageous for porphyrin separation. Analytical LC with its higher flow rate (300 µl/min) in conjunction with a semi-porous C_{18} column media and an almost isocratic gradient allowed much more efficient operation at great resolution. **Figure 5** shows the separation of a standard solution of ZnPPIX (10 pmol on column), MPIX and PPIX (5 pmol on column each).

4.4 Procedure

All experiments were performed in accordance with the declaration of Helsinki and by approval of the Ethics Committee of the Ärztekammer Westfalen-Lippe (2017-169-f-S).

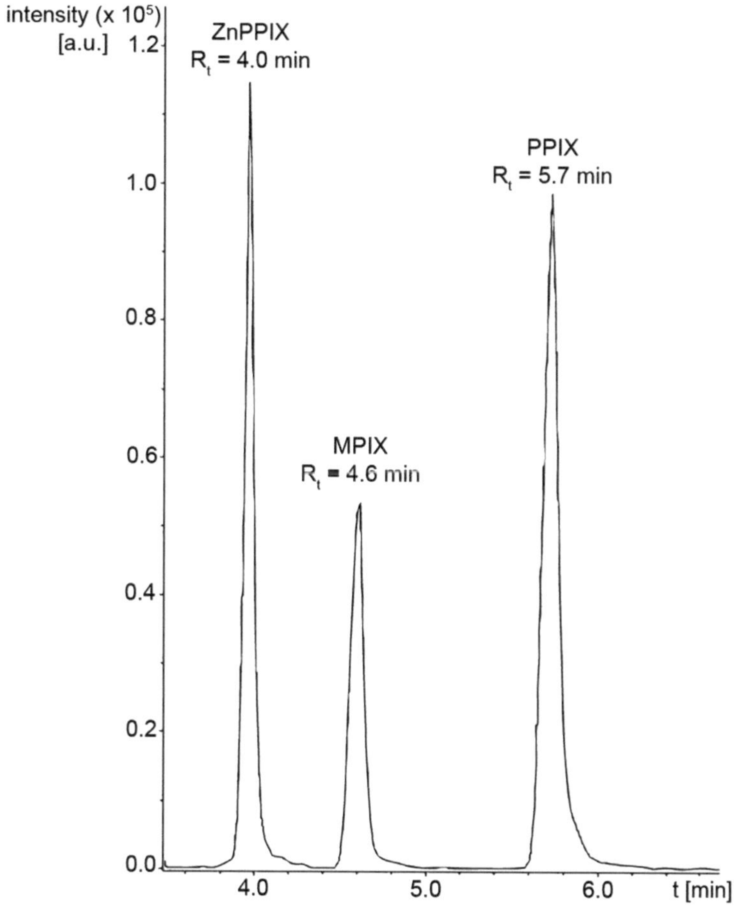

Figure 5.
Extracted ion chromatogram (EIC) of porphyrin standard (ZnPPIX m/z 565.2; MPIX m/z 449.3; PPIX m/z 445.3) as detected with a time-segmented method on the Esquire 3000 ion trap (for parameters see 4.4).

PPIX was obtained from Enzo life sciences GmbH (Lörrach, Germany), MPIX and ZnPPIX from Merck KGaA (Darmstadt, Germany). ACN, water, FA and methanol (MeOH) were all LC-MS grade and purchased from Merck KGaA as were ammonium hydroxide solution and dimethyl sulfoxide (DMSO). SPE cartridges came from Restek GmbH (Bad Homburg, Germany).

For drying of samples, a SpeedVac system (Savant SPD 111V SpeedVac concentrator with vapor trap Savant RVT 5105) was used (Thermo Fisher Scientific, Schwerte, Germany). For centrifugation, a Universal 320R Hettich centrifuge (Tuttlingen, Germany) was utilized.

Hemolysis of 200 µl whole blood or serum was performed by adding 800 µl of water and shaking at room temperature for 1 h. Protein precipitation and porphyrin extraction was achieved by adding 2 ml of ACN, shaking at room temperature for 1 h and centrifuging at 14.000 rcf at 20 °C for 30 min. The clear supernatant was then transferred to an SPE cartridge, which was treated as described in **Table 1**. The 1st and 2nd eluates were dried and finally reconstituted in 15 µl DMSO for HPLC-MS (for parameters see **Tables 2** and **3**).

Raw data were converted using the msConvert toolkit from ProteoWizard software (version 3; [53]). MPIX and PPIX were quantified using the total areas for the three most abundant fragment ions in MS^3 mode (MPIX: extracted ion chromatogram (EIC) *m/z* 449.3, 479.3, 493.3; PPIX: *m/z* 445.3, 459.3, 489.3) and Skyline software (version 20.1; [54]).

Step	Solvent
Cartridge conditioning	2 ml ACN
Equilibration	2 ml water
Sample load	porphyrin extract (~3 ml)
1st Wash	2 ml 5% ammonium hydroxide solution
2nd Wash	2 ml MeOH
1st Elution	2 ml ACN with 2% FA
2nd Elution	2 ml ACN with 20% FA

Table 1.
Protocol for the purification of neutral porphyrin extract with SPE cartridges.

LC-Parameter		
Column	Poroshell C_{18} (2.7 µm, 2.1 mm i.d.) with guard column	
Flow	0.3 ml/min	
Solvent A	95% water/4.9% ACN/0.1% FA	
Solvent B	95% ACN/4.9% water/0.1% FA	
Gradient	t [min]	solvent B [%]
	0.0	70
	1.0	100
	5.0	100
	5.5	70
	10.0	70
Injection volume	10 µl	
MS-Parameter	ESI (+)	
Capillary	−4.5 kV	
End plate	−500 V	
Nebulizer	30.0 psi	
Dry gas	9 l/min	
Dry temperature	320°C	
Scan range	400–700 *m/z*	

Table 2.
HPLC-MS parameter for porphyrin detection.

Analyte	Time [min]	Mode	Isolation *m/z*	Fragmentation amplitude
ZnPPIX	3.4–4.3	MS/MS	625.2 ± 5.0	0.43
MPIX	4.3–5.3	MS/MS	567.3 ± 2.0	0.90
		MS^3	508.3 ± 2.0	0.70
PPIX	5.3–7.0	MS/MS	563.3 ± 2.0	0.85
		MS^3	504.3 ± 2.0	0.70

Table 3.
Parameters for MS/MS detection of ZnPPIX and MS^3 detection of PPIX and MPIX.

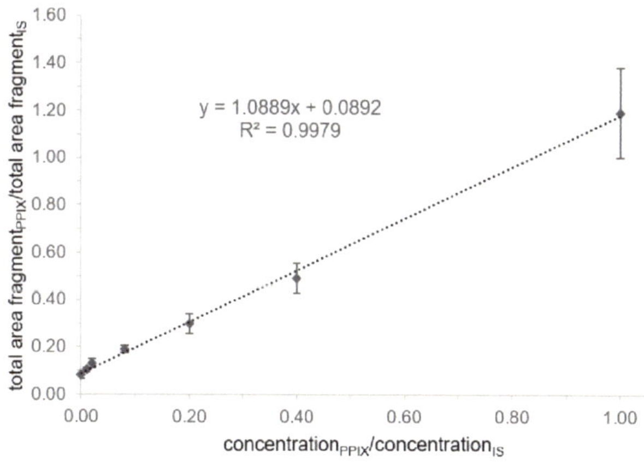

Figure 6.
Calibration curve for PPIX spiked into the whole blood of a healthy volunteer. MPIX (IS) was spiked at 500 fmol/µl. Five replicate injections (10 µl each) were run.

Calibration was performed in the relevant matrix (whole blood of a healthy volunteer) using 500 fmol to 50 pmol PPIX spikes and 500 fmol/µl MPIX (**Figure 6**). The contribution of native PPIX was determined in an aliquot spiked with MPIX only.

5. Measurement of clinical samples

Samples from a patient harboring a GBM and undergoing surgery were collected at different time points before, during and after surgery. The first sample was obtained 1 h prior to ALA administration for the determination of the basic free PPIX level, the following samples were collected 5 and 7 h after ALA administration. Whole blood samples were refrigerated and stored in the dark. Porphyrins were extracted in triplicate using 200 µl of whole blood for each experiment.

Figure 7 shows the results. The level prior to ALA administration was measured twice (two different extractions from the same blood sample) and was 33.9 ±0.4 and 34.9 ± 1.9 pmol PPIX/ml whole blood. After ALA administration, the PPIX level rose about four-fold to 100.9 ± 2.7 and 114.4 ± 11.4 pmol PPIX/ml, respectively, at

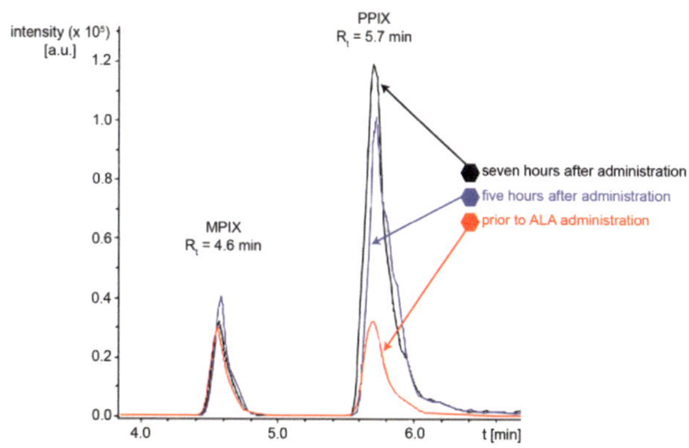

Figure 7.
Overlay of three EIC traces (MPIX m/z 449.3; PPIX m/z 445.3) from measurement of whole blood from a GBM patient taken at different time points. The red line shows a sample prior to ALA administration, whereas the blue and black lines mark the samples 5 and 7 h after ALA administration.

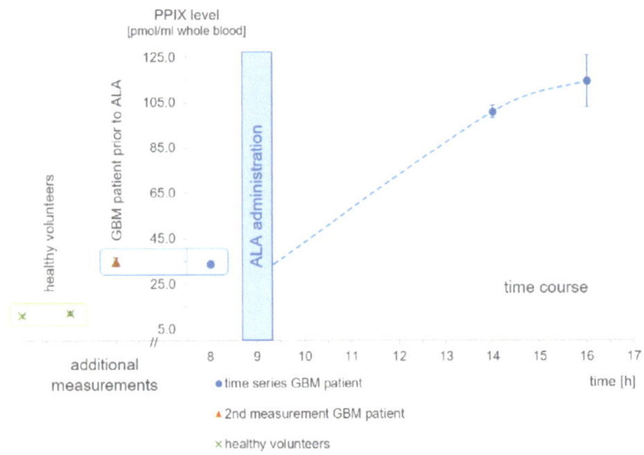

Figure 8.
Time series of a GBM patient with ALA administration at 9 am, illustrated by blue bar and blue points. The pre-ALA PPIX level was confirmed in the same blood sample (orange triangle). PPIX levels in blood of healthy volunteers were determined for comparison (green crosses).

the later time points. For comparison, PPIX levels were determined in two healthy volunteers without ALA administration. They reached only about a third of the pre-surgery level of the patient (11.0 ± 0.7 and 12.1 ± 0.4 pmol PPIX/ml). These observations are summarized in **Figure 8**. So far, having tested one patient only, the results support the hypotheses of elevated PPIX in the circulation as discussed in the literature for other types of cancer [8–11, 32] and the timely and dramatic increase after ALA administration. An extended study involving more probands is planned.

6. Conclusion

MS has the huge advantage over fluorescence-based porphyrin detection that it can pinpoint the individual molecules by their mass and, adding to the specificity, by their fragmentation pattern, which can be generated in the mass spectrometer. Ambiguities as known from fluorescence spectroscopy due to varying or overlapping absorbance maxima and extinction coefficients or fluorescing matrix interferences do not occur; background corrections with complex spectral fitting algorithms are not necessary.

A LC-MS method for the quantification of metal-free PPIX in whole blood was developed. It is short (10 min) and robust (analytical LC, ion trap MS) and provides the necessary specificity and sensitivity. ZnPPIX, MPIX and PPIX can be baseline resolved without the carryover problems observed earlier. The LLE sample preparation provides high extract purity with good recovery. Importantly, ZnPPIX and PPIX can be properly distinguished during SPE clean-up. Matrix effects which would negatively affect HPLC-MS analysis were not observed.

The method is applicable to serum in the same manner, however, serum and plasma PPIX levels are much lower than those of whole blood. Unfortunately, the values reported in the literature lack confidence (**Table 4**). There is no reference range for PPIX in serum or whole blood for healthy individuals. PPIX plasma concentrations in healthy subjects after ALA administration were low and erratic, ranging from below the limit of quantification to hundreds of nmol/l [32, 55, 56]. Often, PPIX was not detected in plasma samples at all [55]. In serum, PPIX levels are lower still, even after ALA administration.

Our results indicate the same extract purity for spiked PPIX extracted from serum as from whole blood. The recovery of PPIX was even slightly better.

	PPIX in plasma (HPLC-FLD) [nmol/l]	Reference
Protoporphyria	2-15	[56]
Healthy adults after ALA administration	0.2-2.8	[32]
	17.8–444.3	[55]

Table 4.
Reported PPIX levels in plasma, measured with HPLC-FLD.

Nevertheless, quantification requires more effort, because only trace amounts of endogenous PPIX were detected in sera of healthy adults so far.

In proof-of-principle experiments the LC-MS method was applied to blood samples from a GBM patient, which confirmed both the elevated PPIX levels in the blood of GBM patients and the increase following ALA administration. The method is now ready for patient screening. Additional resolution and sensitivity for high throughput analysis could be achieved with instrumentation such as a triple quadrupole mass spectrometer, if desired.

Acknowledgements

The authors thank Doreen Ackermann for technical and Carl Zeiss Meditec for financial support. The publication was supported by the Open Access Fonds of the University of Münster.

Conflict of interest

The authors declare no conflict of interest.

Author details

Anna Walke[1,2], Eric Suero Molina[2], Walter Stummer[2] and Simone König[1*]

1 Core Unit Proteomics, Interdisciplinary Center for Clinical Research, University of Münster, Münster, Germany

2 Department of Neurosurgery, University Hospital Münster, Münster, Germany

*Address all correspondence to: koenigs@uni-muenster.de

References

[1] Sachar M, Anderson KE, Ma X. Protoporphyrin IX: The good, the bad, and the ugly. J Pharmacol Exp Ther. 2016;356:267-275. DOI: 10.1124/jpet.115.228130

[2] Mochizuki N, Tanaka R, Grimm B, Masuda T, Moulin M, Smith AG, Tanaka A, Terry MJ. The cell biology of tetrapyrroles: A life and death struggle. Trends Plant Sci. 2010;15:488-498. DOI: 10.1016/j.tplants.2010.05.012

[3] Immenschuh S, Vijayan V, Janciauskiene S, Gueler F. Heme as a target for therapeutic interventions. Front Pharmacol. 2017;8:1-15. DOI: 10.3389/fphar.2017.00146

[4] Ordway GA, Garry DJ. Myoglobin: An essential hemoprotein in striated muscle. J Exp Biol. 2004;207:3441-3446. DOI: 10.1242/jeb.01172

[5] Poulos TL. Heme enzyme structure and function. Chem Rev. 2014;114: 3919-3962. DOI: 10.1021/cr400415k

[6] Stepp H, Stummer W. 5-ALA in the management of malignant glioma. Lasers Surg Med. 2018;50:399-419. DOI: 10.1002/lsm.22933

[7] Stummer W, Pichlmeier U, Meinel T, Wiestler OD, Zanella F, Reulen H-J. Fluorescence-guided surgery with 5-aminolevulinic acid for resection of malignant glioma: A randomised controlled multicentre phase III trial. Lancet Oncol. 2006;7:392-401. DOI: 10.1016/S1470-2045(06)70665-9

[8] Oliveira Silva FR de, Bellini MH, Tristão VR, Schor N, Vieira ND, Courrol LC. Intrinsic fluorescence of protoporphyrin IX from blood samples can yield information on the growth of prostate tumours. J Fluoresc. 2010;20:1159-1165. DOI: 10.1007/s10895-010-0662-9

[9] Kalaivani R, Masilamani V, Sivaji K, Elangovan M, Selvaraj V, Balamurugan SG, Al-Salhi MS. Fluorescence spectra of blood components for breast cancer diagnosis. Photomed Laser Surg. 2008;26:251-256. DOI: 10.1089/pho.2007.2162

[10] Courrol LC, Oliveira Silva FR de, Coutinho EL, Piccoli MF, Mansano RD, Vieira Júnior ND, Schor N, Bellini MH. Study of blood porphyrin spectral profile for diagnosis of tumor progression. J Fluoresc. 2007;17:289-292. DOI: 10.1007/s10895-007-0171-7

[11] Lualdi M, Cavalleri A, Battaglia L, Colombo A, Garrone G, Morelli D, Pignoli E, Sottotetti E, Leo E. Early detection of colorectal adenocarcinoma: A clinical decision support tool based on plasma porphyrin accumulation and risk factors. BMC Cancer. 2018;18:841-849. DOI: 10.1186/s12885-018-4754-2

[12] Lamola AA, Yamane T. Zinc protoporphyrin in the erythrocytes of patients with lead intoxication and iron deficiency anemia. Science. 1974;186:936-938. DOI: 10.1126/science.186.4167.936

[13] Clinical and Laboratory Standards Institute. Erythrocyte protoporphyrin testing: Approved guideline. National Committee for Clinical Laboratory Standards. 1996. C42-A = vol. 16, no. 8. p. 1-40

[14] Centers for Disease Control. A statement by the Centers for Disease Control. CDC Report no. 00-2629. J Pediatr. 1978;93:709-720. DOI: 10.1016/S0022-3476(78)80923-8

[15] Centers for Disease Control. Preventing lead poisoning in young children. A statement by the Centers for Disease Control. 1991:1-104

[16] Balwani M, Desnick RJ. The porphyrias: Advances in diagnosis and treatment. Blood. 2012;120:4496-4504. DOI: 10.1182/blood-2012-05-423186

[17] Piomelli S, Davidow B, Guinee VF, Young P, Gay G. A micromethod for free erythrocyte porphyrins: The FEP test. J Lab Clin Med. 1973;81:932-940. DOI: 10.5555/uri:pii:0022214373903156

[18] Piomelli S. Free erythrocyte porphyrins in the detection of undue absorption of Pb and of Fe deficiency. Clin Chem. 1977;23:264-269. DOI: 10.1093/clinchem/23.2.264

[19] Chisolm JJ, Brown DH. Microscale photofluorometric determination of free erythrocyte porphyrin (protoporphyrin IX). Clin Chem. 1975;21:1669-1682. DOI: 10.1093/clinchem/21.11.1669

[20] Chisolm JJ, Brown DH. Micromethod for zinc protoporphyrin in erythrocytes: Including new data on the absorptivity of zinc protoporphyrin and new observations in neonates and sickle cell disease. Biochem Med. 1979;22:214-237. DOI: 10.1016/0006-2944(79)90008-5

[21] Hart D, Piomelli S. Simultaneous quantitation of zinc protoporphyrin and free protoporphyrin in erythrocytes by acetone extraction. Clin Chem. 1981;27:220-222. DOI: 10.1093/clinchem/27.2.220

[22] Doran D, Mitchell DG. Problems in the determination of erythrocyte protoporphyrin by ethyl acetate-acetic acid extraction. Ann Clin Biochem. 1984;21:141-145. DOI: 10.1177/000456328402100212

[23] Jackson KW. Interlaboratory comparison of results of erythrocyte protoporphyrin analysis. Clin Chem. 1978;24:2135-2138. DOI: 10.1093/clinchem/24.12.2135

[24] Blumberg WE, Eisinger J, Lamola AA, Zuckerman DM. The hematofluorometer. Clin Chem. 1977;23:270-274. DOI: 10.1093/clinchem/23.2.270

[25] Vogeser M, Stauch T, Minder E. Results of a pilot external quality assessment study on free protoporphyrin in erythrocytes. Clin Chem Lab Med. 2011;49:1059-1060. DOI: 10.1515/CCLM.2011.177

[26] Hastka J, Lasserre JJ, Schwarzbeck A, Strauch M, Hehlmann R. Washing erythrocytes to remove interferents in measurements of zinc protoporphyrin by front-face hematofluorometry. Clin Chem. 1992;38:2184-2189. DOI: 10.1093/clinchem/38.11.2184

[27] Hennig G, Gruber C, Vogeser M, Stepp H, Dittmar S, Sroka R, Brittenham GM. Dual-wavelength excitation for fluorescence-based quantification of zinc protoporphyrin IX and protoporphyrin IX in whole blood. J Biophotonics. 2014;7:514-524. DOI: 10.1002/jbio.201200228

[28] Hennig G, Homann C, Teksan I, Hasbargen U, Hasmüller S, Holdt LM, Khaled N, Sroka R, Stauch T, Stepp H, Vogeser M, Brittenham GM. Non-invasive detection of iron deficiency by fluorescence measurement of erythrocyte zinc protoporphyrin in the lip. Nat Commun. 2016;7:10776-10783. DOI: 10.1038/ncomms10776

[29] Homann C, Hennig G, Maier F, Stepp H, Holdt LM, Vogeser M, Sroka R, Koletzko B. Non-invasive measurement of erythrocyte zinc protoporphyrin in children. Pediatr Res. 2019;85:349-354. DOI: 10.1038/s41390-018-0247-x

[30] Gotelli GR, Wall JH, Kabra PM, Marton LJ. Simultaneous liquid-chromatographic determination of zinc protoporphyrin IX, protoporphyrin IX, and coproporphyrin in whole blood.

Clin Chem. 1980;26:205-208. DOI: 10.1093/clinchem/26.2.205

[31] Schwartz S, Stephenson B, Sarkar D, Freyholtz H, Ruth G. Quantitative assay of erythrocyte “free” and zinc-protoporphyrin: Clinical and genetic studies. Int J Biochem. 1980;12:1053-1057. DOI: 10.1016/0020-711X(80)90212-8

[32] Ota U, Fukuhara H, Ishizuka M, Abe F, Kawada C, Tamura K, Tanaka T, Inoue K, Ogura S-I, Shuin T. Plasma protoporphyrin IX following administration of 5-aminolevulinic acid as a potential tumor marker. Mol Clin Oncol. 2015;3:797-801. DOI: 10.3892/mco.2015.549

[33] Hindmarsh JT, Oliveras L, Greenway DC. Biochemical differentiation of the porphyrias. Clin Biochem. 1999;32:609-619. DOI: 10.1016/S0009-9120(99)00067-3

[34] Macours P, Cotton F. Improvement in HPLC separation of porphyrin isomers and application to biochemical diagnosis of porphyrias. Clin Chem Lab Med. 2006;44:1433-1440. DOI: 10.1515/CCLM.2006.266

[35] Lim CK, Peters TJ. Urine and faecal porphyrin profiles by reversed-phase high-performance liquid chromatography in the porphyrias. Clin Chim Acta. 1984;139:55-63. DOI: 10.1016/0009-8981(84)90192-X

[36] Lim CK, Li F, Peters TJ. High-performance liquid chromatography of porphyrins. J Chromatogr. 1988;429:123-153. DOI: 10.1016/S0378-4347(00)83869-4

[37] Hindmarsh JT, Oliveras L, Greenway DC. Plasma porphyrins in the porphyrias. Clin Chem. 1999;45:1070-1076

[38] Benton CM, Lim CK, Moniz C, Jones DJL. Ultra high-performance liquid chromatography of porphyrins in clinical materials: Column and mobile phase selection and optimisation. Biomed Chromatogr. 2012;26:714-719. DOI: 10.1002/bmc.1720

[39] Sullivan SA, Streit BR, Ferguson EL, Jean PA, McNett DA, Llames LT, DuBois JL. Mass-spectrometric profiling of porphyrins in complex biological samples with fundamental, toxicological, and pharmacological applications. Anal Biochem. 2015;478:82-89. DOI: 10.1016/j.ab.2015.03.004

[40] Danton M, Lim CK. Porphyrin profiles in blood, urine and faeces by HPLC/electrospray ionization tandem mass spectrometry. Biomed Chromatogr. 2006;20:612-621. DOI: 10.1002/bmc.656

[41] Bu W, Myers N, McCarty JD, O'Neill T, Hollar S, Stetson PL, Sved DW. Simultaneous determination of six urinary porphyrins using liquid chromatography-tandem mass spectrometry. J Chromatogr B. 2003;783:411-423. DOI: 10.1515/9783110887242.3

[42] Ausió X, Grimalt JO, Ozalla D, Herrero C. On-line LC-MS analysis of urinary porphyrins. Anal Chem. 2000;72:4874-4877. DOI: 10.1021/ac0005060

[43] Kandoussi H, Zeng J, Shah K, Paterson P, Santockyte R, Kadiyala P, Shen H, Shipkova P, Langish R, Burrrell R, Easter J, Mariannino T, Marathe P, Lai Y, Zhang Y, Pillutla R. UHPLC-MS/MS bioanalysis of human plasma coproporphyrins as potential biomarkers for organic anion-transporting polypeptide-mediated drug interactions. Bioanalysis. 2018;10:633-644. DOI: 10.4155/bio-2017-0246

[44] Fyrestam J, Bjurshammar N, Paulsson E, Johannsen A, Östman C. Determination of porphyrins in oral

bacteria by liquid chromatography electrospray ionization tandem mass spectrometry. Anal Bioanal Chem. 2015;407:7013-7023. DOI: 10.1007/s00216-015-8864-2

[45] Fyrestam J, Östman C. Determination of heme in microorganisms using HPLC-MS/MS and cobalt(III) protoporphyrin IX inhibition of heme acquisition in *Escherichia coli*. Anal Bioanal Chem. 2017;409:6999-7010. DOI: 10.1007/s00216-017-0610-5

[46] Moiyadi A, Stummer W. δ-Aminolevulinic acid-induced fluorescence-guided resection of brain tumors. Neurol India. 2015;63:155-165. DOI: 10.4103/0028-3886.156275

[47] Stoyanov GS, Dzhenkov D, Ghenev P, Iliev B, Enchev Y, Tonchev AB. Cell biology of glioblastoma multiforme: From basic science to diagnosis and treatment. Med Oncol. 2018;35:27-36. DOI: 10.1007/s12032-018-1083-x

[48] Peters I, König S. Extraction of δ-aminolevulinic acid-induced protoporphyrin IX from blood and serum and measurement using analytical chromatography coupled to mass spectrometry. Mercator J Biomol Anal. 2019;3:6-9. URN: urn:nbn:de:hbz:6-12189602979

[49] Demarque DP, Crotti AEM, Vessecchi R, Lopes JLC, Lopes NP. Fragmentation reactions using electrospray ionization mass spectrometry: An important tool for the structural elucidation and characterization of synthetic and natural products. Nat Prod Rep. 2016;33:432-455. DOI: 10.1039/c5np00073d

[50] Gledhill M. The detection of iron protoporphyrin (heme b) in phytoplankton and marine particulate material by electrospray ionisation mass spectrometry - comparison with diode array detection. Anal Chim Acta. 2014;841:33-43. DOI: 10.1016/j.aca.2014.06.045

[51] Vessecchi R, Crotti AEM, Guaratini T, Colepicolo P, Galembeck S, Lopes N. Radical ion generation processes of organic compounds in electrospray ionization mass spectrometry. Mini-Rev Org Chem. 2007;4:75-87. DOI: 10.2174/157019307779815910

[52] van Berkel GJ, Zhou F. Observation of gas-phase molecular dications formed from neutral organics in solution via the controlled-current electrolytic process inherent to electrospray. J Am Soc Mass Spectrom. 1996;7:157-162. DOI: 10.1016/1044-0305(95)00627-3

[53] Kessner D, Chambers M, Burke R, Agus D, Mallick P. ProteoWizard: Open source software for rapid proteomics tools development. Bioinformatics. 2008;24:2534-2536. DOI: 10.1093/bioinformatics/btn323

[54] Adams KJ, Pratt B, Bose N, Dubois LG, St John-Williams L, Perrott KM, Ky K, Kapahi P, Sharma V, MacCoss MJ, Moseley MA, Colton CA, MacLean BX, Schilling B, Thompson JW. Skyline for small molecules: A unifying software package for quantitative metabolomics. J Proteome Res. 2020;19:1447-1458. DOI: 10.1021/acs.jproteome.9b00640

[55] Dalton JT. Clinical pharmacokinetics of 5-aminolevulinic acid in healthy volunteers and patients at high risk for recurrent bladder cancer. J Pharm Exp Ther. 2002;301:507-512. DOI: 10.1124/jpet.301.2.507

[56] Thomas L. Porphyrine. In: Thomas L, Editor. Labor und Diagnose: Indikation und Bewertung von Laborbefunden für die medizinische Diagnostik. 6nd Ed. Frankfurt/Main: TH-Books-Verlags-Gesellschaft; 2005. p. 646-659. ISBN: 3-9805215-5-9

Chapter 5

Mass Spectrometry in Clinical Laboratories

Jadranka Miletić Vukajlović and Tanja Panić-Janković

Abstract

The analyses performed in clinical laboratories require a high level of precision, selectivity, and sensitivity. The rising number of therapeutic agents from both the field of small and large molecules and the increasing use of modern screening approaches have brought mass spectrometry into almost every clinical laboratory. The need to screen the patients and to follow the therapy's success can often be fulfilled only by the highly selective and sensitive targeted approach with mass spectrometry. With improving instrument design and miniaturization of the separation technologies, mass spectrometry is no longer an exotic analytical approach. The use of mass spectrometry is now not restricted to the use in a clinical laboratory, but it is used in operating rooms for instant and on-site helping the surgeons with defining the margin of the tissue to be extracted. In this manuscript, we describe the use of mass spectrometry for selected clinical applications and show the possible way of future applications.

Keywords: Clinical laboratory, antibiotics, newborn screening, mass spectrometry

1. Introduction

The use of mass spectrometry in the clinical laboratory has become a standard for analysis of different substances such as antibiotics, for newborn screening, detection of immune-suppressive drugs, or the analysis of therapeutic antibodies used for the treatment of different diseases.

The focus of the use of mass spectrometry in clinical settings is the analysis of clinical samples and monitoring levels of active compounds and their metabolites in patients' blood and urine samples. The high sensitivity and specificity of the mass spectrometer and the possibility to perform specific detection of target analytes by applying MRM/SRM (multiple reaction monitoring/selected reaction monitoring) enable a targeted and highly specific analytical approach. The methods developed need a separation method in front of the MS and several companies such as Chromsystems (https://www.chromsystems.com/), ThermoFisher Scientific (https://www.thermofisher.com/at/en/home/clinical/diagnostic-testing/clinical-chemistry-drug-toxicology-testing/therapeutic-drug-monitoring.html) or BioRad (www.bio-rad.com), to name just a few, have developed fully verified and certified analytical systems. The interested reader is encouraged to search the internet for additional providers and systems.

Applying chromatography and mass spectrometry has its primary values in relatively fast detection and measuring of multiple analytes in a single sample with high sensitivity and high selectivity. In clinical routine, the key challenge for identifying and analyzing active compounds is having the sensitivity of the analytical system needed and required to detect and quantify low-concentration analytes.

One of the challenges for using the MS in a clinical laboratory was the low ion yield, which significantly hampered the development of clinical applications. However, the development of new analytical systems, especially of new ion inlets and ion funnel designs with the most widely used electrospray ionization (ESI) sources has significantly improved ion focusing and ion transfer, which, finally, resulted in the overall increased sensitivity.

The quality of electrospray is highly dependable on separation conditions, i.e. mobile phase, presence or absence of salts, flow speed, column's inner diameter, etc. In proteomics, the use of columns with 50 µm or 75 µm ID is state-of-the-art. However, the columns operated at a low flow rate of several hundreds of nanoliters/ minute are still rare in clinical analysis although they can provide a significant increase in analysis's sensitivity.

However, currently, the use of nanoflow separation still cannot cope with the demand for high sample throughput in clinical applications. Currently, the closest compromise between sensitivity and throughput is the use of the microbore and capillary columns of 300 µm – 500 µm and 1 mm – 2 mm inner diameter.

A new and exciting application of mass spectrometry in the clinical environment is the use of "live-MS" during surgical operations. Further development of this approach will revolutionize the diagnostics and help surgeons in extracting e.g. tumors with higher accuracy and higher yield.

2. Clinical applications

2.1 Analysis of antibiotics

Antibiotics, either cytotoxic or cytostatic to the microorganisms, have been widely used to treat and prevent infectious diseases and allow the body's natural defenses to eliminate them. They usually have a role to inhibit the synthesis of proteins, deoxyribonucleic acid (DNA), ribonucleic acid (RNA), or other specific actions [1]. Using the energy-dependent transport mechanisms in ribosomal sites, antibiotics target bacterial cell wall by attaching to them, which consequently results in inhibition of protein synthesis and subsequent cell death [2].

With the discovery of penicillin by Sir Alexander Fleming, a new, modern, chapter of innovation and antibiotics development began [3]. Today, there are different classes of antibiotics (**Table 1**) and they are widely used not only in human medicine but also in veterinary medicine and aquaculture [35]. However, antibiotics inadvertently released into the environment can cause a massive threat to the ecosystems and subsequently to human health. Consequently, they may accumulate in food and, which is much more worrying, antibiotic resistance of human pathogens might develop through the transfer of environmental bacteria genes (ARGs) [36–40]. Furthermore, sensitive individuals might experience allergic reactions triggered by antibiotic residues in food. Furthermore, the ingestion of subtherapeutic doses of antibiotics and uncontrolled use of antibiotics may initiate the development of drug-resistant strains of bacteria that initially appeared in hospitals only, where most antibiotics were being used [41].

In recent years, more and more scientific data and news reporting the misuse and the overuse of antibiotics [42, 43], the environment exposure pathways [44, 45], and the presence of antibiotic-resistant strains [46, 47] became available.

Over the years, numerous analytical methods have been developed and described to determine antibiotic residues in the environment and food.

Different classes of antibiotics	References
β-lactam	[4–6]
Sulfonamides	[7–9]
Aminoglycosides	[10–12]
Tetracyclines	[13, 14]
Chloramphenicol	[15, 16]
Macrolides	[17–19]
Glycopeptides	[20–22]
Oxazolidinones	[23, 24]
Ansamycins	[25, 26]
Quinolones	[27, 28]
Streptogramins	[29, 30]
Lipopeptides	[31, 32]
Antibiotic Resistance	[33, 34]

Table 1.
Overview of different classes of antibiotics and some related references describing their mass spectrometry analysis.

Chromatographic separation and detection of antibiotics and their metabolites using various detectors is the most widely used analytical approach for monitoring and determination.

The electrospray ionization technique has become the technique of choice in many areas of analyzing biologically relevant macromolecules [48]. The soft ionization MS techniques - matrix-assisted laser desorption/ionization (MALDI) and ESI [49, 50] proved to be the best approach for analysis due to the efficient ionization of polar antibiotics. Depending on the ionization mode applied, it has been shown that most antibiotics yield a better signal when positive ionization is used, with the most commonly formed protonated molecular ion $[M + H]^+$. Determination of analytes trace levels in complex biological matrices using the molecular ion generally is not enough selective due to the limited resolution of unit-mass MS instruments. Therefore, these obstacles are overcome by using modern equipment consisting of liquid chromatography (LC) MS instrumentation with tandem MS (MS/MS), which became the technique of choice in quantitative bioanalysis. Tandem MS capacities enhance selectivity and signal-to-noise ratio and provide essential structural information based on which it is possible to identify the structural conformation of analyzed samples. For these reasons, many laboratories use triple quadrupole (Q) MS/MS over ion trap (IT) MS instruments in routine practice for detection and analysis of antibiotics and other drug residues [51–55]. This advantage is reflected in its quantitative features regarding IT MS with its MS^n capabilities, which are highly beneficial for analysis of analyte's molecular structure and identification.

With technological advancement, instruments providing accurate-mass, high-resolution (HR) time-of-flight (TOF) MS, single TOF-MS, or hybrid instruments combined with a quadrupole (Q-TOF-MS) and the collision cell for MS/MS analysis became available. HR-MS has entered every day's practice of clinical laboratories as a viable alternative to traditional triple quadrupole mass spectrometer. The versatility of HR-MS (especially hybrid HR-MS) is reflected in increased selectivity by eliminating potential interferences originating from the matrix with remarkably similar mass-to-charge ratio (m/z) as of the measured analytes, but with a different structure.

Unlike IT and Q MS, TOF-MS is a pulsed, and a non-scanning MS. TOF-MS can acquire full spectral data, thus separates and detects ions of various m/z by measuring the time taken for the ions to travel through a field-free region. Therefore, these instruments are mostly combined with a fast LC separation if used for rapid non-targeted screening [56].

Based on all the above-mentioned, LC–MS is an essential factor in the pharmaceutical industry and clinical laboratory due to the possibility of identifying impurities in synthetic products, characterize metabolites, and perform quantitative bioanalysis.

The following section of this chapter provides an overview of examples of mass spectrometry usage in clinical laboratories for detection and characterization of antibiotics in a different sample including pharmaceutical, blood (plasma, serum), environmental water samples (waste, surface, and drinking water) [57–62], animal and plants and products of animal and plant origin, etc. [63–65].

Depending on sample matrices such as muscle, liver, kidney, egg, milk, or honey, multiclass methods based on LC–MS or LC–MS/MS are used for the analysis of antibiotics residues [66–69]. The complexity of the methods depends also on the complexity of the sample preparation. Therefore, screening methods try to avoid complicated sample preparation such as solid-phase extraction (SPE) and the evaporation of the purified extract before the chromatographic separation whereas quantitative methods do not bypass this step [70]. Chico et al. [71] analyzed 39 analytes residues that belong to 5 families of antibiotics with different physicochemical properties which include sulfonamides (SAs), quinolones (Qs), tetracyclines (TCs), macrolides (MCs), and penicillins (PCs). To shorten the analysis time, their method set-up was based on ultra-high-pressure liquid chromatography (UHPLC), like in Yamaguchi et al. [72] and Tian et al. [73], combined with tandem mass spectrometry–MS/MS with ESI. Mass spectrometry parameters were determined and optimized by an infusion of standard solutions to accomplish the highest sensitivity. The singly protonated molecular ion was selected and used as the precursor ion for all compounds $[M + H]^+$, and the cone voltage was adjusted to its maximum signal at the first quadrupole of the mass spectrometer. The success of this method proved to be exceptional and, for that reason, was introduced as a method at the laboratory of Agència de la Salut Pública de Barcelona.

Several analytical methods are currently available to separately detect the fluoroquinolone and sulfonamide classes of antibiotics in manure, surface water, wastewater, and groundwater [74–76]. Haller et al. [77] focused on liquid–liquid extraction followed by LC–MS analysis of veterinary antibiotics (sulfonamides and trimethoprim), which are most commonly being leaked into the aquatic environment. Based on published LC–MS methods for sulfonamides separation on a reversed-phase chromatographic column, ammonium acetate buffered water and acetonitrile were used as mobile phases. The most successful baseline separation was achieved using the buffered mobile phases at pH 4.6, which enables more stable retention times and better peak shapes for almost all analyzed analytes due to their pKa. Analytes appear to be more hydrophobic and retain better on an RP HPLC column. Haller et al. [77] acquired SIM mass spectra of all samples (antibiotics and of the internal standard) in the full scan mode, using positive and negative electrospray ionization. Single-protonated $[M + H]^+$ or the $[M–H]^-$, and several (two to three) additional fragments that were generated through the in-source fragmentation, which is typical for single quadrupole mass spectrometer and that yielded the best signal-to-noise (S/N) ratios were selected for confirmation. The advantage of this method is multiple: a very simple extraction process was applied, thus sample preparation is faster, the method does not require tandem mass spectrometry, the method is capable of detecting the investigated pharmaceuticals, to determine the half-lives of antibiotics in manure slurry, and to establish mass balances from antibiotic contents in medicinal feed to quantities.

Renew et al. [74] analyzed groups of antibiotics (fluoroquinolones, sulfonamides, and trimethoprim) simultaneously at sub micrograms per liter concentrations in wastewater effluents using readily available LC–MS techniques. Quantification and identification were performed by applying fluorescence detection and additionally confirmed by tandem LC–MS.

Following Chico et al. SPE followed by LC–MS analysis is utilized by this method. Hirsch et al. [78] and Hartig et al. [79] developed LC–MS techniques for sulfonamides detection. The application of this method allowed preliminary determination of the occurrence of these antibiotics in municipal wastewater treatment plants. Usually, normal phase chromatography (NPC), which implied that the use of a polar stationary phase, was used for the LC separation. However, the NP stationary phases usually show large heterogeneity, which was also observed in this experiment as a consequence of peak tailing and non-linear retention factors with varying analyte concentrations [80]. Different solvents were used to accomplish elution in NPC, from non-polar organic to some variants like the use of isohydric solvents [81, 82]. Some obstacles like lack of retention of highly hydrophilic compounds with ionizable functional groups have been exceeded by ion-exchange chromatography [83] or ion pairing on reversed-phase (RP) columns [84]. However, for those analytes with high hydrophilicity the problem has been overcome using hydrophilic interaction chromatography (HILIC). In contrast to the RP LC, the gradient elution in HILIC starts with a low-polarity, low acquoeus organic solvent and elutes polar analytes by increasing the polar content. In addition, in HILIC, ion pair reagents are not required, and the separation system can be easily coupled to MS, especially in the ESI mode [85].

A large topic opens when it comes to antibiotic treatment, as well as establishing resistance to them. MALDI-TOF [33, 86] technique is an ionization technique that allows the analysis of biomolecules and is used to monitor antibiotic treatment as well as rapid detection of antibiotic resistance. The feasibility of MALDI-TOF MS identification of bacterial colonies from solid media has been evaluated on a wide range of clinically relevant bacterial strains as well as yeast isolates.

MALDI-TOF MS whole-cell extracts identification represents a new method for obtaining a characteristic bacterial fingerprint, which allows for distinction of microorganisms based on different genera, species, and from different strains of the same species. The advantages of using this method are numerous: identification can be achieved in a short time after culture isolation, sensitivity is high, ability to detect microorganisms is not limited to prespecified targets, mass spectra obtained for unknown microorganisms are compared with reference database to achieve the identification. Therefore, MALDI-TOF MS represents a reliable method for rapid bacteria and fungi identification in a clinical setting.

The biggest global challenge due to growth rates of multi-drug-resistant microorganisms, especially in hospital settings, introduces new analytical methods not only for prevention and treatment but also for the detection and determination of antibiotic-resistant species. Numerous MALDI-TOF MS-based methods have been recommended for the rapid detection of antibiotic-resistance in bacterial pathogens isolated from bloodstream infections as well as for detection of antimicrobial-resistance in pathogenic fungi. Methods based on an assessment of β-lactamase activity, biomarkers detection responsible for drug-resistance, and/or non-susceptibility, and the comparison of bacteria proteomic profiles incubated with or without antimicrobial drugs, are the most widely studied [33].

2.2 Newborn screening-amino acid analysis

The newborn screening (NBS) program was developed for early diagnosis of asymptomatic newborns at risk for rare diseases such as inborn errors of metabolism (IEM). However, meanwhile, the screening program includes all newborn independently of the risk. IEM is a serious, degenerative, chronic disease with painful and unpredictable clinical manifestations varying from apparent clinical state or obfuscated with other diseases' symptoms to differing degrees of mental retardation and physical disability [87]. These diseases often result in disturbed levels of amino acids or acylcarnitines, which are used as diagnostic markers for IEM. Many problems correlated to irregular amino acid metabolism generate abnormal ammonia concentrations, resulting in an increased turnover of amino acids for energy production or an indicator of alterations in urea cycle metabolism [88]. In the 1960s, the first NBS for the most frequent aminoacidopathia, which is phenylketonuria (PKU), (Guthrie & Susi, 1963), was developed using a dried blood spot (DBS) [89]. It was established to detect PKU and enable early treatment and prevent neurodevelopmental problems if untreated. From the newborn screening perspective, time is a vital factor in the disease etiology. Without screening, many disorders cannot be recognized on time and untreated patients can exhibit serious symptoms of the disease and end up in a coma or even face death. Children diagnosed on time and with adequate treatment are functional, have reducing sequelae or at least substantially lessening organ damages, and may live normal life [90].

Thirty years ago, the first report was published using tandem mass spectrometry (TMS) [91] for analyzing multiple acylcarnitines and amino acids on a single blood spot. The following development of TMS had been introduced as combined with fast atom bombardment (FAB) and electrospray ionization (ESI) [92–96] recently with high-resolution liquid or gas chromatography (LC, GC) respectively [97], and direct analysis in real-time [98] mass spectrometry for newborn screening purposes. According to multiple authors [99–103], blood spot extracts are analyzed by FIA coupled to triple quadrupole (TQ) TMS. Although TQ instruments possess robustness and sensitivity, these instruments also experience monoisotopic interferences with naturally occurring 13C isotopologues, in-source fragmentation interferences, and low mass resolving power, which leads to difficulties separating isobaric compounds with identical quantifying product ions.

The possibility of multiple disorders detection in a single blood spot shortly after birth increased with new technologies in mass spectrometry. TMS is the most widely used instrument for the detection and analysis of amino acids in the DBS and represents one of the most important advancements in the neonatal screening approach [104, 105]. **Figure 1** shows a general scheme of DMB sample preparation for MS analysis.

Analysis of specific amino acids proved to be adequate indicator for the presence of certain disorders in newborns. By measuring fluctuations and disturbances in amino acid metabolism, a diverse group of disorders can be identified and confirmed [107]. Disorders that affect the metabolism of amino acids include PKU, tyrosinemia type I (TYR I), maple syrup urine disease (MSUD), homocystinuria (HCY), argininosuccinic aciduria (ASA), and citrullinemia (CIT) (**Table 2**). These disorders are autosomal recessive and can be confirmed by analyzing amino acid concentrations in body fluids. Because of more than 500 confirmed disorders detected, the use of TMS for clinical screening in a newborn is the method of choice for a few million newborn screenings worldwide [90, 106, 112, 114–116].

When it comes to the operation mode of the instrument, TMS can be operated in different modes such as neutral loss scanning, precursor ion scanning, and multiple reaction monitoring. When neutral loss scanning mode is used, all

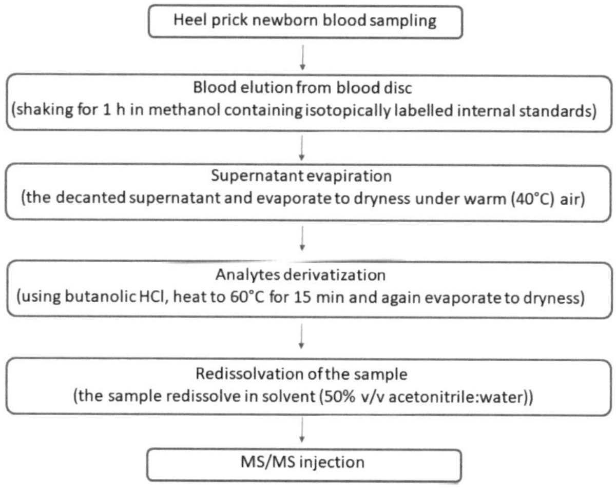

Figure 1.
Example of DMB sample preparation steps for MS analysis [106].

Disorders	Marker(s)	Method	References
Classic phenylketonuria (PKU)	Phe, Tyr	LC–MS/MS	[108]
Tyrosinemia type I	SA, Tyr	LC–MS/MS	[109]
Marple syrup urine disease (MSUD)	Leu, Ile, Val	LC–MS/MS	[110]
Homocystinuria (HCY)	Met	MALDI-TOF MS	[111]
Argininosuccinic aciduria (ASA)	Asa, Cit	HPLC–MS/MS	[112]
Citrullinemia (type I and II) (CIT)	Cit	ESI-MS/MS	[113]

Phe (phenylalanine), Tyr (tyrosine), SA (succinylacetone), Leu (leucine), Ile (isoleucine), Val (valine), Met (methionine), Asa (argininosuccinic acid), Cit (citrulline).

Table 2.
Exemplary overview of the parameters used in newborn screening and the technology applied.

precursors undergoing the loss of a common, neutral, fragment such as water, ammonia, or a phosphate-group are being detected and can be used for further experiments. The neutral loss method is applied for the detection of amino acids due to the neutral loss of m/z 46, which is being shared by many amino acids during fragmentation [117].

Acylcarnitines are detected by this method, as they produce a characteristic fragment ion of 85 m/z. On the other hand, other amino acids, e.g. arginine, ornithine, and citrulline, split off other fragments during collision-induced dissociation. Due to their basic functional group that fragments easily, the most common loss is a combination of butyl formate and ammonia [105, 116–119]. Therefore, for the detection of all amino acids, acylcarnitines, and other biological compounds in NBS, it is safer and better to use the multiple-reaction-monitoring mode for data acquisition (MRM).

To detect the compound of interest, this method requires individual mass transition optimization to achieve the highest selectivity and sensitivity for both amino acids and acylcarnitines, and only selected amino acids can be measured quantitatively and selectively [116]. It is crucial to emphasize the difference between "screening" and quantification in TMS analysis of amino acids. For the diagnosis of PKU, determination of phenylalanine/tyrosine (Phe/Tyr) is of higher importance than a precise measurement of only phenylalanine [120, 121].

For transient neonatal hypertyrosinemia, an elevated level of tyrosine is usually detected by TMS. To differentiate, the diagnosis of TYR I can be established by detecting the presence of succinylacetone in serum or urine [122]. It is important to emphasize that the high concentration of tyrosine is not always a companion of TYR I [123]. Allard et al. developed a method for verifying TYR1 by using succinylacetone as a determination marker (SUAC) in DBS [124]. Some data reported that this method is unmistakably sensitive and specific, while other reports pointed out that false-positive results were also obtained [125]. Many screening programs for homocystinuria have combined determination of methionine (Met) as a primary marker, methionine, and phenylalanine ratio (Met/Phe), and the total homocysteine (tHcy) as a second-tier marker in DBS [126, 127].

Bartl et al. incorporated the LC–MS/MS analysis as a potential first step in screening clinically symptomatic high-risk populations for the two types of HCY and severe B-vitamin deficiencies. In several IEMs, increased reactive oxygen species (ROS) causes pathophysiological oxidative damage that, in the case of HCY, excess Hcy directly supports ROS formation in the form of O_2^-, hydrogen radical, or H_2O_2 [128]. Elevated Hcy concentration is deemed a risk factor for neurodegenerative diseases inducing neurological dysfunction via oxidative stress [129]. Mild to moderate increases in Hcy levels have been associated with both vascular dementia and Alzheimer's disease (AD) [130, 131] and with a possible increased risk of developing Parkinson's disease at a later age [132–134].

Many cases on the diagnosis of PKU [135], MSUD [136], and HCY [126] in newborn blood spots using amino acid analysis by FAB TMS were also reported.

Screening of a large number of disorders was established when Rashed et al. [137] used ESI for analyzing butyl esters of amino acids and acylcarnitines. Consequently, clinical laboratories around the world use this automated sample insertion and data analysis method for a newborn screening procedure to detect and analyze selected amino acids and acylcarnitines [138]. Chace et al. [139] first described the use of TMS for MSUD NBS and recommended the determination of total leucine (Xle) in combination with a total leucine and phenylalanine ratio (Xle/Phe, respectively) for improved detection. In the following studies, recommendations for MSUD detection was based on an elevated Xle or leucine (Leu) [112, 140–147]. Some studies reported that Val is also required for referral [148–150] while others did report Val, but without the cut-off value [139, 143, 151]. Other studies also included the Xle/alanine (Ala) ratio [151]. In a long 11-years-long study in the Netherlands, MSUD NBS was measured in almost two million newborns using TMS, and MSUD was confirmed for 4 patients and 118 false-positive referrals. The authors recommended Xle/Phe ratio as a promising additional marker ratio to their MSUD NBS strategy and advised consideration of method implementation in the Dutch NBS program [138].

Although sensitive, the newborn screening does have some limitations, and therefore, particular caution is required to the common symptoms that may indicate a metabolic disorder. Its goal is to prevent morbidity and mortality through the early detection of metabolic disorders. A significant number of these disorders may present in the neonatal period; therefore, the need for a newborn screening technique is rising. Tandem mass spectrometry has emerged rapidly in previous years as a crucial multiplex testing technique for biochemical genetics analysis and newborn screening and the number of possible disorders that may be included for NBS has exponentially increased.

2.3 On-site mass spectrometry in OP-room

The continuous increase in the prevalence of cancer requires continuous innovation of both diagnostics and treatment. One of the crucial steps in cancer therapy is as complete as possible surgical removal of the tumor from the surrounding healthy tissue. This so-called negative margin assessment is of critical importance for complete tumor removal and for achieving tumor remission and improve the overall survival rate of patients.

Surgical on-site decision-making could be enhanced with devices and different methods that give an instant and adequate biochemical information about the multiple biopsies or continuous sampling during surgery. Different MS platforms have shown to be able to provide and substantial impact in surgical decision-making process in different points during clinical workflow. To achieve this goal, surgeons would greatly benefit from using mass spectroscopy during the actual operation is going and having immediate information about the resected tumor. This would significantly increase the rate of successfully and almost completely removed tumors and reduce the risk of tumor recurrence. One of the main requirements, or a minimum requirement, for surgery, is that the selected technique delivers fast and accurate information on unprocessed samples and that the ionization is performed as ambient ionization thus eliminating the need for suction and minimizing the use of other solvents than sterile water. DESI (desorption electrospray ionization) was the first technology to be used for the offline analysis of resected tissue. For DESI, a spray of charged solvents is directed onto the tissue's surface and secondary droplets containing the analytes are desorbed and sampled by the MS. Based on this approach, Eberlin et al. [152, 153] developed the MasSpec Pen for intraoperative MS analyses and rapid diagnostics of cancer.

One example is the use of the MasSpec Pen (MS Pen) for diagnostics of ovarian cancer [154] published by Sans et al. Ovarian cancer is a highly lethal disease that is very often diagnosed very late and it is the fifth leading cause of deaths among women [155, 156]. Furthermore, as with other cancers, accurate diagnosis is of extreme importance for the selection of the treatment and development of precision medicine approach and personalized medicine and therapy. For ovarian cancer, two therapy scenarios are possible: a) cytoreductive surgery before chemotherapy and b) surgery upon chemotherapy for tumors that cannot be fully resected. The timing for the cytoreductive surgery is of great importance and in both cases, it is very important to differentiate the tumor from the healthy tissue with high precision. Identification of a tumor can also be very difficult in cases where scarring or some other fibrous tissue is present and, sometimes, healthy tissue is removed, which should be avoided. Unlike iKnife, the MS Pen uses a water droplet to extract molecules from the tissue [157] and transfer it to the ion source. The full process is very fast, it needs no derivatization or other kinds of sample preparation and the acting surgeon gets an instant result based on a database search, which can help to properly identify the resected tissue and enable better determination of the resection margin. Sans et al. [154] have described the use of MassSpecPen for rapid diagnosis of ovarian cancer. The authors analyzed tissue samples from the tissue bank or from prospectively collected samples from endometriosis surgeries to establish the database needed. The authors analyzed the presence of small metabolites such as glycerophosphoinositol, glycerophosphoserine, glutathione, and glycerophospholipid. It was found that normal ovarian tissue was characterized by presence of ascorbate and some other small metabolites with a relatively high abundance in comparison to cancer samples.

The iKnife was developed with the same purpose as the MassSpec Pen but it relies on ionizing analytes in the smoke plum that is generated during electro cauterization of the tissue during the surgery [158]. Unlike the MassSpec Pen, iKnife is preferably used to identify lipids in the smoke plum. By comparing the mass spectra of the sample generated during the surgery and the database that was established earlier, the surgeon sees the result instantly on the screen and can make decisions about further procedure. St. John et al. described the use of iKnife for the identification of breast pathology for breast cancer surgery [159]. The aerosol produced by the monopolar hand piece used in surgery was aspirated and analytes therein were ionized in the mass spectrometer's ion source. Generated data were used to identify the tissue by applying multivariate analysis. The method proved to be able to identify the substances within a very short time range of 1.8 seconds. Here, the spectral differences that arise between the two operational modes of the electrosurgical knife – the "cut" and the "clog" – were combined to create a multivariate statistical model and to allow for using both modes during the surgery.

A further application where the iKnife was applied is ex-vivo use for diagnosis of cervical disease. The specimen obtained by cervical punch biopsy can either bee snap frozen and used for confirmation of the conventional histology analysis or it can be analyzed immediately upon sampling. Tzafetas et al. [158] was showed that the application of this technology enabled identification of lipids that characterize cancer, the normal tissue, and samples affected by HPV.

MALDI mass spectrometry has already proved efficient for analyzing microorganism and for the offline imaging (MSI) tissue analyses. It is the MSI that represent an encouraging tool to support histopathology analyses and the decision-making processes. MALDI MSI captures the entire spectrum of biomolecules, including specific biomarkers, providing enhanced discriminating power over the visual inspection of tissue and placing it as a proper assisting method in diagnosis procedure.

With the progress of ambient mass spectrometry techniques, such DESI, MS became a powerful methodology for characterizing lipids within tumor specimens. The DESI MS analysis can be performed with minimal sample preparation and it provides molecular information from tissue samples rapidly. This qualifies the DESI and MALDI methods as a diagnostic method in the OP room. In addition to tumor classification, defining tumor subtypes, and identifying tumor grade, this method also provides necrotic tumor tissue identification, an indicator of high-grade malignancy, and can help distinguish necrotic tumor tissue from viable tumor regions [159–162].

3. Conclusion

The use of MS in clinical laboratories worldwide increasing, and, as a result, substantial improvements in assay performance are occurring rapidly in many areas such as toxicology, endocrinology, and biochemical genetics. Numerous types of mass spectrometers are being used for the characterization of small molecules such as drugs of abuse, steroids, amines, amino and organic acids, as well as larger compounds such as proteins and ribosomal RNA.

The development of MS technologies has pushed clinical MS toward the analysis of peptides and proteins for diagnostic examination. However, the quantitative analysis of proteins by MS is still a challenging area of laboratory medicine, which faces many challenges before being fit for a routine application. Also, MS contributes to the quality of the many test results (standardization of assays for steroids, lipids, hemoglobin A1c, etc.), and is used as a standard method in all US states for newborn screening. Furthermore, it is important to address that nearly every institution sends tests to the reference laboratories which frequently perform these tests using MS. With the improved functionality that benefits novel front-end modifications and computational abilities, MS can now be used for nontraditional clinical analyses, including clinical microbiology applications for bacteria differentiation and in surgical operating rooms.

We did not address the role of MALDI imaging technology for application in pathology, but it is one of the fastest-growing application of mass spectrometry in clinical settings and the growth is only impeded by a lack of fast and easy to use software packages for fast identification of analytes others than peptides or lipids.

Conflict of interest

The authors declare no conflict of interest.

Author details

Jadranka Miletić Vukajlović[1]* and Tanja Panić-Janković[2]

1 Department of Physical Chemistry, "VINČA" Institute of Nuclear Sciences - National Institute of the Republic of Serbia, University of Belgrade, Belgrade, Serbia

2 Clinical Institute of Laboratory Medicine, Medical University of Vienna, Vienna, Austria

*Address all correspondence to: jadranka@vin.bg.ac.rs

References

[1] Levy SB, Marshall B: Antibacterial resistance worldwide: causes, challenges, and responses. Nature Medicine. 2004,10:122-129. DOI: 10.1038/nm1145

[2] Maranan MC, Moreira B, Boyle-Vavra S, et al.: Antimicrobial resistance in staphylococci: Epidemiology, molecular mechanisms, and clinical relevance. Infectious Disease Clinics of North America. 1997;11:813-849. DOI: 10.1016/S0891-5520(05)70392-5

[3] Fleming A. On the antibacterial action of cultures of a penicillium, with special reference to their use in the isolation of B. influenzae. The British journal of experimental pathology. 1929;10:223-236.

[4] Petrović M, Hernando MD, Díaz-Cruz MS, Barceló D. Liquid chromatography–tandem mass spectrometry for the analysis of pharmaceutical residues in environmental samples: a review. Journal of Chromatography A. 2005 Mar 4;1067(1-2):1-4.

[5] Mastovska K, Lightfield AR. Streamlining methodology for the multiresidue analysis of β-lactam antibiotics in bovine kidney using liquid chromatography–tandem mass spectrometry. Journal of Chromatography A. 2008 Aug 22;1202(2):118-123.

[6] Yu J, Liu J, Li Y, Yu J, Zhu W, Liu Y, Shen L. Rapid detection of carbapenemase activity of Enterobacteriaceae isolated from positive blood cultures by MALDI-TOF MS. Annals of clinical microbiology and antimicrobials. 2018 Dec 1;17(1):22.

[7] Białk-Bielińska A, Kumirska J, Palavinskas R, Stepnowski P. Optimization of multiple reaction monitoring mode for the trace analysis of veterinary sulfonamides by LC–MS/MS. Talanta. 2009 Dec 15;80(2):947-953.

[8] García-Galán MJ, Villagrasa M, Díaz-Cruz MS, Barceló D. LC-QqLIT MS analysis of nine sulfonamides and one of their acetylated metabolites in the Llobregat River basin. Quantitative determination and qualitative evaluation by IDA experiments. Analytical and bioanalytical chemistry. 2010 Jun 1;397(3):1325-1334.

[9] Yu W, Li K, Liu Z, Zhang H, Jin X. Novelty aqueous two-phase extraction system based on ionic liquid for determination of sulfonamides in blood coupled with high-performance liquid chromatography. Microchemical Journal. 2018 Jan 1;136:263-269.

[10] Oertel R, Neumeister V, Kirch W. Hydrophilic interaction chromatography combined with tandem-mass spectrometry to determine six aminoglycosides in serum. Journal of chromatography A. 2004 Nov 26;1058(1-2):197-201.

[11] Kumar P, Rubies A, Companyo R, Centrich F. Hydrophilic interaction chromatography for the analysis of aminoglycosides. Journal of separation science. 2012 Feb;35(4):498-504.

[12] Sharma D, Kumar B, Lata M, Joshi B, Venkatesan K, Shukla S, Bisht D. Comparative proteomic analysis of aminoglycosides resistant and susceptible Mycobacterium tuberculosis clinical isolates for exploring potential drug targets. PLoS One. 2015 Oct 5;10(10):e0139414.

[13] Meisen I, Wisholzer S, Soltwisch J, Dreisewerd K, Mormann M, Müthing J, Karch H, Friedrich AW. Normal silica gel and reversed phase thin-layer chromatography coupled with UV spectroscopy and IR-MALDI-o-TOF-MS

for the detection of tetracycline antibiotics. Analytical and bioanalytical chemistry. 2010 Dec 1;398(7-8):2821-2831.

[14] Gavilán RE, Nebot C, Miranda JM, Martín-Gómez Y, Vázquez-Belda B, Franco CM, Cepeda A. Analysis of tetracyclines in medicated feed for food animal production by HPLC-MS/MS. Antibiotics. 2016 Mar;5(1):1.

[15] Boyd B, Björk H, Billing J, Shimelis O, Axelsson S, Leonora M, Yilmaz E. Development of an improved method for trace analysis of chloramphenicol using molecularly imprinted polymers. Journal of Chromatography A. 2007 Dec 7;1174(1-2):63-71.

[16] Rezende DR, Filho NF, Rocha GL. Simultaneous determination of chloramphenicol and florfenicol in liquid milk, milk powder and bovine muscle by LC–MS/MS. Food Additives & Contaminants: Part A. 2012 Apr 1;29(4):559-570.

[17] Göbel A, McArdell CS, Suter MJ, Giger W. Trace determination of macrolide and sulfonamide antimicrobials, a human sulfonamide metabolite, and trimethoprim in wastewater using liquid chromatography coupled to electrospray tandem mass spectrometry. Analytical Chemistry. 2004 Aug 15;76(16):4756-4764.

[18] Berrada H, Borrull F, Font G, Moltó JC, Marcé RM. Validation of a confirmatory method for the determination of macrolides in liver and kidney animal tissues in accordance with the European Union regulation 2002/657/EC. Journal of Chromatography A. 2007 Jul 20;1157(1-2):281-288.

[19] Wang J. Analysis of macrolide antibiotics, using liquid chromatography-mass spectrometry, in food, biological and environmental matrices. Mass spectrometry reviews. 2009 Jan;28(1):50-92.

[20] Lin YS, Tsai PJ, Weng MF, Chen YC. Affinity capture using vancomycin-bound magnetic nanoparticles for the MALDI-MS analysis of bacteria. Analytical chemistry. 2005 Mar 15;77(6):1753-1760.

[21] König K, Kobold U, Fink G, Leinenbach A, Dülffer T, Thiele R, Zander J, Vogeser M. Quantification of vancomycin in human serum by LC-MS/MS. Clinical Chemistry and Laboratory Medicine (CCLM). 2013 Sep 1;51(9):1761-1769.

[22] Oyaert M, Peersman N, Kieffer D, Deiteren K, Smits A, Allegaert K, Spriet I, Van Eldere J, Verhaegen J, Vermeersch P, Pauwels S. Novel LC–MS/MS method for plasma vancomycin: Comparison with immunoassays and clinical impact. Clinica Chimica Acta. 2015 Feb 20;441:63-70.

[23] Ji HY, Lee HW, Chang SG, Lee JJ, Rhee JK, Kim WB, Lee HS. Liquid chromatography-tandem mass spectrometry for the determination of a new oxazolidinone antibiotic DA-7867 in human plasma. Biomedical Chromatography. 2004 Mar;18(2):86-89.

[24] Wang Y, Li X, Fu Y, Chen Y, Wang Y, Ye D, Wang C, Hu X, Zhou L, Du J, Shen J. Association of florfenicol residues with the abundance of oxazolidinone resistance genes in livestock manures. Journal of Hazardous Materials. 2020 May 28:123059.

[25] Jensen MR, Schoepfer J, Radimerski T, Massey A, Guy CT, Brueggen J, Quadt C, Buckler A, Cozens R, Drysdale MJ, Garcia-Echeverria C. NVP-AUY922: a small molecule HSP90 inhibitor with potent antitumor activity in preclinical

breast cancer models. Breast Cancer Research. 2008 Apr 1;10(2):R33.

[26] Liu LL, Chen ZF, Liu Y, Tang D, Gao HH, Zhang Q, Gao JM. Molecular networking-based for the target discovery of potent antiproliferative polycyclic macrolactam ansamycins from Streptomyces cacaoi subsp. asoensis. Organic Chemistry Frontiers. 2020;7(24):4008-4018.

[27] Jansen LJ, Bolck YJ, Rademaker J, Zuidema T, Berendsen BJ. The analysis of tetracyclines, quinolones, macrolides, lincosamides, pleuromutilins, and sulfonamides in chicken feathers using UHPLC-MS/MS in order to monitor antibiotic use in the poultry sector. Analytical and Bioanalytical Chemistry. 2017 Aug 1;409(21):4927-4941.

[28] Zhao Y, Tang M, Liu F, Li H, Wang H, Xu D. Highly Integrated Microfluidic Chip Coupled to Mass Spectrometry for Online Analysis of Residual Quinolones in Milk. Analytical chemistry. 2019 Sep 30;91(21):13418-13426.

[29] Vrioni G, Tsiamis C, Oikonomidis G, Theodoridou K, Kapsimali V, Tsakris A. MALDI-TOF mass spectrometry technology for detecting biomarkers of antimicrobial resistance: current achievements and future perspectives. Annals of Translational Medicine. 2018 Jun;6(12).

[30] Civitareale C, Mestria S, Gallo P, Giannetti L, Neri B, Stacchini P, Fiori M. Multi-drug ultraperformance liquid chromatography/tandem mass spectrometry method to quantify antimicrobials in feeding stuffs at carry-over level. Rapid Communications in Mass Spectrometry. 2018 Nov 15;32(21):1831-1842.

[31] Armbruster KM, Meredith TC. Enrichment of bacterial lipoproteins and preparation of N-terminal lipopeptides for structural determination by mass spectrometry. JoVE (Journal of Visualized Experiments). 2018 May 21(135):e56842.

[32] Liu Y, Teng K, Wang T, Dong E, Zhang M, Tao Y, Zhong J. Antimicrobial Bacillus velezensis HC6: production of three kinds of lipopeptides and biocontrol potential in maize. Journal of Applied Microbiology. 2020 Jan;128(1):242-254.

[33] Idelevich EA, Sparbier K, Kostrzewa M, Becker K. Rapid detection of antibiotic resistance by MALDI-TOF mass spectrometry using a novel direct-on-target microdroplet growth assay. Clinical Microbiology and Infection. 2018;24:738-743.DOI: 10.1016/j.cmi.2017.10.016

[34] Angeletti S. Matrix assisted laser desorption time of flight mass spectrometry (MALDI-TOF MS) in clinical microbiology. Journal of microbiological methods. 2017 Jul 1;138:20-29.

[35] Mastovska, Katerina. "Multiresidue analysis of antibiotics in food of animal origin using liquid chromatography–mass spectrometry." Mass spectrometry in food safety. Humana Press, 2011. 267-307. DOI: 10.1007/978-1-61779-136-9_12

[36] Bengtsson-Palme J, Larsson DJ. Antibiotic resistance genes in the environment: prioritizing risks. Nature Reviews Microbiology. 2015;13:396. DOI: 10.1038/nrmicro3399-c1

[37] Du L, Liu W. Occurrence, fate, and ecotoxicity of antibiotics in agro-ecosystems. A review. Agronomy for sustainable development. 2012;32:309-327. DOI: 10.1007/s13593-011-0062-9

[38] Li B, Yang Y, Ma L, Ju F, Guo F, Tiedje JM, Zhang T. Metagenomic and network analysis reveal wide distribution and co-occurrence of

environmental antibiotic resistance genes. The ISME journal. 2015;9:2490-2502. DOI: 10.1038/ismej.2015.59

[39] Martínez JL, Coque TM, Baquero F. What is a resistance gene? Ranking risk in resistomes. Nature Reviews Microbiology. 2015;13:116-123. DOI: 10.1038/nrmicro3399

[40] Van Boeckel TP, Brower C, Gilbert M, Grenfell BT, Levin SA, Robinson TP, Teillant A, Laxminarayan R. Global trends in antimicrobial use in food animals. Proceedings of the National Academy of Sciences. 2015;112:5649-5654. DOI: 10.1073/pnas.1503141112

[41] Yang Y, Song W, Lin H, Wang W, Du L, Xing W. Antibiotics and antibiotic resistance genes in global lakes: A review and meta-analysis. Environment international. 2018;116:60-73. DOI: 10.1016/j.envint.2018.04.011

[42] Chang Y, Chusri S, Sangthong R, McNeil E, Hu J, Du W, Li D, Fan X, Zhou H, Chongsuvivatwong V, Tang L. Clinical pattern of antibiotic overuse and misuse in primary healthcare hospitals in the southwest of China. PloS one. 2019;14:e0214779. DOI: 10.1371/journal.pone.0214779

[43] Limmathurotsakul D, Sandoe JA, Barrett DC, Corley M, Hsu LY, Mendelson M, Collignon P, Laxminarayan R, Peacock SJ, Howard P. 'Antibiotic footprint' as a communication tool to aid reduction of antibiotic consumption. Journal of Antimicrobial Chemotherapy. 2019 Aug 1;74(8):2122-2127. DOI: 10.1093/jac/dkz185

[44] Ben Y, Fu C, Hu M, Liu L, Wong MH, Zheng C. Human health risk assessment of antibiotic resistance associated with antibiotic residues in the environment: a review. Environmental research. 2019;169:483-493. DOI: 10.1016/j.envres.2018.11.040

[45] Singer AC, Shaw H, Rhodes V, Hart A. Review of antimicrobial resistance in the environment and its relevance to environmental regulators. Frontiers in microbiology. 2016 Nov 1;7:1728. DOI: 10.3389/fmicb.2016.01728

[46] Bergeron S, Boopathy R, Nathaniel R, Corbin A, LaFleur G. Presence of antibiotic resistant bacteria and antibiotic resistance genes in raw source water and treated drinking water. International Biodeterioration & Biodegradation. 2015;102:370-374. DOI: 10.1016/j.ibiod.2015.04.017

[47] Ljubojević D, Puvača N, Pelić M, Todorović D, Pajić M, Milanov D, Velhner M. Epidemiological significance of poultry litter for spreading the antibiotic-resistant strains of Escherichia coli. World's Poultry Science Journal. 2016;72:485-494. DOI: 10.1017/S004393391600043X

[48] Yi X, Bayen S, Kelly BC, Li X, Zhou Z. Improved detection of multiple environmental antibiotics through an optimized sample extraction strategy in liquid chromatography-mass spectrometry analysis. Analytical and bioanalytical chemistry. 2015;407:9071-9083. DOI: 10.1007/s00216-015-9074-7

[49] Wu C, Kim HK, van Wezel GP, Choi YH. Metabolomics in the natural products field–a gateway to novel antibiotics. Drug Discovery Today: Technologies. 2015;13:11-17. DOI: 10.1016/j.ddtec.2015.01.004

[50] Jamshidi-Aidji M, Morlock GE. Bioprofiling of unknown antibiotics in herbal extracts: Development of a streamlined direct bioautography using Bacillus subtilis linked to mass spectrometry. Journal of Chromatography A. 2015;1420:110-118. DOI: 10.1016/j.chroma.2015.09.061

[51] Moldoveanu S, Scott W, Zhu J. Analysis of small carbohydrates in several bioactive botanicals by gas

chromatography with mass spectrometry and liquid chromatography with tandem mass spectrometry. Journal of separation science. 2015;38:3677-3686. DOI: 10.1002/jssc.201500573

[52] Huang Z, Pan XD, Huang BF, Xu JJ, Wang ML, Ren YP. Determination of 15 β-lactam antibiotics in pork muscle by matrix solid-phase dispersion extraction (MSPD) and ultra-high pressure liquid chromatography tandem mass spectrometry. Food control. 2016;66:145-150. DOI: 10.1016/j.foodcont.2016.01.037

[53] Gentili A, Perret D, Marchese S. Liquid chromatography-tandem mass spectrometry for performing confirmatory analysis of veterinary drugs in animal-food products. TrAC Trends in Analytical Chemistry. 2005;24:704-733. DOI: 10.1016/j.trac.2005.02.007

[54] Blasco C, Picó Y, Torres CM. Progress in analysis of residual antibacterials in food. TrAC Trends in Analytical Chemistry. 2007;26:895-913. DOI: 10.1016/j.trac.2007.08.001

[55] Panić-Janković T, Mitulović G. Human chorionic gonadotrophin pharmaceutical formulations of urinary origin display high levels of contaminant proteins–A label-free quantitation proteomics study. Electrophoresis. 2019;40(11):1622-1629. DOI: 10.1002/elps.201900087

[56] Mastovska, K. Recent developments in chromatographic techniques, in Comprehensive Analytical Chemistry. In: Pico, Y, editor. Food Contaminants and Residue Analysis. 1st ed. Oxford: Elsevier; 2008. p. 175-200. DOI: 10.1007/978-1-61779-136-9_12

[57] Samanidou VF, Hapeshi EA, Papadoyannis IN. Rapid and sensitive high-performance liquid chromatographic determination of four cephalosporin antibiotics in pharmaceuticals and body fluids. Journal of Chromatography B. 2003;788:147-158. DOI: 10.1016/S1570-0232(02)01040-1

[58] Samanidou VF, Ioannou AS, Papadoyannis IN. The use of a monolithic column to improve the simultaneous determination of four cephalosporin antibiotics in pharmaceuticals and body fluids by HPLC after solid phase extraction—a comparison with a conventional reversed-phase silica-based column. Journal of Chromatography B. 2004;809:175-182. DOI: 10.1016/j.jchromb.2004.06.019

[59] Kim C, Ryu HD, Chung EG, Kim Y, Lee JK. A review of analytical procedures for the simultaneous determination of medically important veterinary antibiotics in environmental water: Sample preparation, liquid chromatography, and mass spectrometry. Journal of environmental management. 2018;217:629-645. DOI: 10.1016/j.jenvman.2018.04.006

[60] Ding J, Ren N, Chen L, Ding L. On-line coupling of solid-phase extraction to liquid chromatography–tandem mass spectrometry for the determination of macrolide antibiotics in environmental water. Analytica Chimica Acta. 2009;634:215-221. DOI: 10.1016/j.aca.2008.12.022

[61] Kim C, Ryu HD, Chung EG, Kim Y. Determination of 18 veterinary antibiotics in environmental water using high-performance liquid chromatography-q-orbitrap combined with on-line solid-phase extraction. Journal of Chromatography B. 2018;1084:158-165. DOI: 10.1016/j.jchromb.2018.03.038

[62] Gracia-Lor E, Sancho JV, Hernández F. Multi-class determination of around 50 pharmaceuticals, including 26 antibiotics, in environmental and

wastewater samples by ultra-high performance liquid chromatography–tandem mass spectrometry. Journal of Chromatography A. 2011;1218:2264-2275. DOI: 10.1016/j.chroma.2011.02.026

[63] Vishnuraj MR, Kandeepan G, Rao KH, Chand S, Kumbhar V. Occurrence, public health hazards and detection methods of antibiotic residues in foods of animal origin: A comprehensive review. Cogent Food & Agriculture. 2016;2:1235458. DOI: 10.1080/23311932.2016.1235458

[64] Wegener HC. Antibiotics in animal feed and their role in resistance development. Current opinion in microbiology. 2003;6:439-445. DOI: 10.1016/j.mib.2003.09.009

[65] Gousia P, Economou V, Bozidis P, Papadopoulou C. Vancomycin-resistance phenotypes, vancomycin-resistance genes, and resistance to antibiotics of enterococci isolated from food of animal origin. Foodborne Pathogens and Disease. 2015;12:214-220. DOI: 10.1089/fpd.2014.1832

[66] Rosén J. Efficient and sensitive screening and confirmation of residues of selected polyether ionophore antibiotics in liver and eggs by liquid chromatography-electrospray tandem mass spectrometry. Analyst. 2001;126:1990-1995.

[67] Matabudul DK, Conway B, Lumley I, Sumar S. The simultaneous determination of the ionophore antibiotics in animal tissues and eggs by tandem electrospray LC–MS–MS. Food Chemistry. 2001;75:345-354. DOI: 10.1016/S0308-8146(01)00174-1

[68] Holstege DM, Puschner B, Whitehead G, Galey FD. Screening and mass spectral confirmation of β-lactam antibiotic residues in milk using LC-MS/MS. Journal of agricultural and food chemistry. 2002;50:406-411. DOI: 10.1021/jf010994s

[69] Debayle D, Dessalces G, Grenier-Loustalot MF. Multi-residue analysis of traces of pesticides and antibiotics in honey by HPLC-MS-MS. Analytical and Bioanalytical Chemistry. 2008;391:1011-1020. DOI: 10.1007/s00216-008-2003-2

[70] Denev P, Ciz M, Ambrozova G, Lojek A, Yanakieva I, Kratchanova M. Solid-phase extraction of berries' anthocyanins and evaluation of their antioxidative properties. Food Chemistry. 2010;123:1055-1061. DOI: 10.1016/j.foodchem.2010.05.061

[71] Chico J, Rúbies A, Centrich F, Companyó R, Prat MD, Granados M. High-throughput multiclass method for antibiotic residue analysis by liquid chromatography–tandem mass spectrometry. Journal of Chromatography A. 2008;1213:189-199. DOI: 10.1016/j.chroma.2008.10.023

[72] Yamaguchi T, Okihashi M, Harada K, Uchida K, Konishi Y, Kajimura K, Hirata K, Yamamoto Y. Rapid and easy multiresidue method for the analysis of antibiotics in meats by ultrahigh-performance liquid chromatography–tandem mass spectrometry. Journal of agricultural and food chemistry. 2015;63:5133-5140. DOI: 10.1021/acs.jafc.5b00170

[73] Tian H, Wang J, Zhang Y, Li S, Jiang J, Tao D, Zheng N. Quantitative multiresidue analysis of antibiotics in milk and milk powder by ultra-performance liquid chromatography coupled to tandem quadrupole mass spectrometry. Journal of chromatography B. 2016;1033:172-179. DOI: 10.1016/j.jchromb.2016.08.023

[74] Renew JE, Huang CH. Simultaneous determination of fluoroquinolone, sulfonamide, and trimethoprim antibiotics in wastewater using tandem solid phase extraction and liquid chromatography–electrospray mass spectrometry. Journal of

Chromatography A. 2004;1042:113-121. DOI: 10.1016/j.chroma.2004.05.056

[75] Adrian J, Pinacho DG, Granier B, Diserens JM, Sánchez-Baeza F, Marco MP. A multianalyte ELISA for immunochemical screening of sulfonamide, fluoroquinolone and ß-lactam antibiotics in milk samples using class-selective bioreceptors. Analytical and bioanalytical chemistry. 2008;391:1703-1712. DOI: 10.1007/s00216-008-2106-9

[76] Heller DN, Nochetto CB, Rummel NG, Thomas MH. Development of multiclass methods for drug residues in eggs: hydrophilic solid-phase extraction cleanup and liquid chromatography/tandem mass spectrometry analysis of tetracycline, fluoroquinolone, sulfonamide, and β-lactam residues. Journal of agricultural and food chemistry. 2006;54:5267-5278. DOI: 10.1021/jf0605502

[77] Haller MY, Müller SR, McArdell CS, Alder AC, Suter MJ. Quantification of veterinary antibiotics (sulfonamides and trimethoprim) in animal manure by liquid chromatography–mass spectrometry. Journal of Chromatography A. 2002;952:111-120. DOI: 10.1016/S0021-9673(02)00083-3

[78] Hirsch R, Ternes TA, Haberer K, Mehlich A, Ballwanz F, Kratz KL. Determination of antibiotics in different water compartments via liquid chromatography–electrospray tandem mass spectrometry. Journal of chromatography A. 1998;815:213-223. DOI: 10.1016/S0021-9673(98)00335-5

[79] Hartig C, Storm T, Jekel M. Detection and identification of sulphonamide drugs in municipal waste water by liquid chromatography coupled with electrospray ionisation tandem mass spectrometry. Journal of Chromatography A. 1999;854:163-173. DOI: 10.1016/S0021-9673(99)00378-7

[80] Bernal J, Ares AM, Pól J, Wiedmer SK. Hydrophilic interaction liquid chromatography in food analysis. Journal of Chromatography A. 2011;1218:7438-7452. DOI: 10.1016/j.chroma.2011.05.004

[81] Tóth G, Panić-Janković T, Mitulović G. Pillar array columns for peptide separations in nanoscale reversed-phase chromatography. Journal of Chromatography A. 2019;1603:426-432. DOI: 10.1016/j.chroma.2019.06.067.

[82] Bens GA, Crombez E, Moerloose PD. Applications of a modified "Isohydric solvent system" in HPLC on silica gel for the analysis of the macrolide antibiotics turimycins and spiramycins. Journal of Liquid Chromatography. 1982 Jan 1;5(8):1449-1465. DOI: 10.1080/01483918208062843

[83] Serrano JM, Silva M. Rapid and sensitive determination of aminoglycoside antibiotics in water samples using a strong cation-exchange chromatography non-derivatisation method with chemiluminescence detection. Journal of Chromatography A. 2006;1117:176-183. DOI: 10.1016/j.chroma.2006.03.086

[84] Kim BH, Lee SC, Lee HJ, Ok JH. Reversed-phase liquid chromatographic method for the analysis of aminoglycoside antibiotics using pre-column derivatization with phenylisocyanate. Biomedical Chromatography. 2003;17:396-403. DOI: 10.1002/bmc.254

[85] Kahsay G, Song H, Van Schepdael A, Cabooter D, Adams E. Hydrophilic interaction chromatography (HILIC) in the analysis of antibiotics. Journal of pharmaceutical and biomedical analysis. 2014;87:142-154. DOI: 10.1016/j.jpba.2013.04.015

[86] Kostrzewa M, Sparbier K, Maier T, Schubert S. MALDI-TOF MS: an upcoming tool for rapid detection of

antibiotic resistance in microorganisms. PROTEOMICS–Clinical Applications. 2013;7:767-778. DOI: 10.1002/prca.201300042

[87] Adhikari AN, Gallagher RC, Wang Y, Currier RJ, Amatuni G, Bassaganyas L, Chen F, Kundu K, Kvale M, Mooney SD, Nussbaum RL: The role of exome sequencing in newborn screening for inborn errors of metabolism. Nature Medicine. 2020;10:1-6. DOI: 10.1038/s41591-020-0966-5

[88] Vanessa Shaw. Dixon M, MacDonald A, White FJ. Disorders of Amino Acid Metabolism, Organic Acidaemias and Urea Cycle Disorders. In: Vanessa Shaw, editor. Clinical Paediatric Dietetics, 5th ed. Wiley-Blackwell; 2020. p. 513-598 DOI: 10.1002/9781119467205.ch28

[89] Guthrie R, Susi A. A simple phenylalanine method for detecting phenylketonuria in large populations of newborn infants. Pediatrics. 1963;32:338-343

[90] Céspedes N, Valencia A, Echeverry CA, Arce-Plata MI, Colón C, Castiñeiras DE, Hurtado PM, Cocho JA, Herrera S, Arévalo-Herrera M. Reference values of amino acids, acylcarnitines and succinylacetone by tandem mass spectrometry for use in newborn screening in southwest Colombia. Colombia Médica. 2017; 48:113-9. DOI: 10.25100/cm.v48i3.2180

[91] Millington DS, Kodo N, Norwood DL, Roe CR. Tandem mass spectrometry: a new method for acylcarnitine profiling with potential for neonatal screening for inborn errors of metabolism. Journal of inherited metabolic disease. 1990;13:321-324. DOI: 10.1007/BF01799385

[92] Chace DH, DiPerna JC, Naylor EW. Laboratory integration and utilization of tandem mass spectrometry in neonatal screening: a model for clinical mass spectrometry in the next millennium. Acta Paediatrica. 1999;88:45-47. DOI: 10.1111/j.1651-2227.1999.tb01156.x

[93] Chen WH, Hsieh SL, Hsu KP, Chen HP, Su XY, Tseng YJ, Chien YH, Hwu WL, Lai F. Web-based newborn screening system for metabolic diseases: machine learning versus clinicians. Journal of medical Internet research. 2013;15:e98. DOI: 10.2196/jmir.2495

[94] Hall PL, Marquardt G, McHugh DM, Currier RJ, Tang H, Stoway SD, Rinaldo P. Postanalytical tools improve performance of newborn screening by tandem mass spectrometry. Genetics in Medicine. 2014;16:889-895. DOI: 10.1038/gim.2014.62

[95] la Marca G. Mass spectrometry in clinical chemistry: the case of newborn screening. Journal of pharmaceutical and biomedical analysis. 2014;101:174-182. DOI: 10.1016/j.jpba.2014.03.047

[96] Clinical and Laboratory Standard Institute. Newborn screening by tandem mass spectrometry; Approved guideline. Wayne (PA): Clinical and Laboratory Standard Institute, 2010.

[97] Miller IV JH, Poston PA, Karnes HT. A quantitative method for acylcarnitines and amino acids using high resolution chromatography and tandem mass spectrometry in newborn screening dried blood spot analysis. Journal of Chromatography B. 2012;903:142-149. DOI: 10.1016/j.jchromb.2012.07.008

[98] Wang C, Zhu H, Cai Z, Song F, Liu Z, Liu S. Newborn screening of phenylketonuria using direct analysis in real time (DART) mass spectrometry. Analytical and bioanalytical chemistry. 2013;405:3159-3164. DOI: 10.1007/s00216-013-6713-8

[99] Chaimbault PE, Petritis KO, Elfakir CL, Dreux MI. Determination of 20 underivatized proteinic amino acids by ion-pairing chromatography and pneumatically assisted electrospray mass spectrometry. Journal of Chromatography A. 1999;855:191-202. DOI: 10.1016/S0021-9673(99)00685-8

[100] Petritis K, Chaimbault P, Elfakir C, Dreux M. Parameter optimization for the analysis of underivatized protein amino acids by liquid chromatography and ionspray tandem mass spectrometry. Journal of Chromato graphy A. 2000;896:253-263. DOI: 10.1016/S0021-9673(00)00582-3

[101] Piraud M, Vianey-Saban C, Petritis K, Elfakir C, Steghens JP, Morla A, Bouchu D. ESI-MS/MS analysis of underivatised amino acids: a new tool for the diagnosis of inherited disorders of amino acid metabolism. Fragmentation study of 79 molecules of biological interest in positive and negative ionisation mode. Rapid Communications in Mass Spectrometry. 2003;17:1297-1311. DOI: 10.1002/rcm.1054

[102] Piraud M, Vianey-Saban C, Petritis K, Elfakir C, Steghens JP, Bouchu D. Ion-pairing reversed-phase liquid chromatography/electrospray ionization mass spectrometric analysis of 76 underivatized amino acids of biological interest: a new tool for the diagnosis of inherited disorders of amino acid metabolism. Rapid Communications in Mass Spectrometry: An International Journal Devoted to the Rapid Dissemination of Up-to-the-Minute Research in Mass Spectrometry. 2005;19:1587-602. DOI: 10.1002/rcm.1957

[103] Zakaria R, Allen KJ, Koplin JJ, Roche P, Greaves RF. Advantages and challenges of dried blood spot analysis by mass spectrometry across the total testing process. Ejifcc. 2016;27:288.

[104] Chace DH, Kalas TA, Naylor EW. Use of tandem mass spectrometry for multianalyte screening of dried blood specimens from newborns. Clinical chemistry. 2003;49:1797-1817. DOI: 10.1373/clinchem.2003.022178

[105] Fabie NA, Pappas KB, Feldman GL. The current state of newborn screening in the United States. Pediatric Clinics. 2019;66:369-386. DOI: 10.1016/j.pcl.2018.12.007

[106] Wiley V, Carpenter K, Wilcken B. Newborn screening with tandem mass spectrometry: 12 months' experience in NSW Australia. Acta Paediatrica. 1999;88:48-51.DOI: 10.1111/j.1651-2227.1999.tb01157.x

[107] Pieragostino D, Cicalini I, Di Michele S, Fusilli P, Cotugno G, Ferrante R, Bucci I, Dionisi-Vici C, Stuppia L, De Laurenzi V, Rossi C. A Case of Suspected Hyperphenyla laninemia at Newborn Screening by Tandem Mass Spectrometry during Total Parenteral Nutrition. Metabolites. 2020;10:44. DOI: 10.3390/metabo10020044

[108] Gouda AS, Nazim WS. Development of a simple method for the analysis of phenylalanine in dried blood spot using tandem mass spectrometry. Egyptian Journal of Medical Human Genetics. 2020;21(1):1-6.

[109] la Marca G, Malvagia S, Pasquini E, Innocenti M, Fernandez MR, Donati MA, Zammarchi E. The inclusion of succinylacetone as marker for tyrosinemia type I in expanded newborn screening programs. Rapid Communications in Mass Spectrometry: An International Journal Devoted to the Rapid Dissemination of Up-to-the-Minute Research in Mass Spectrometry. 2008;22(6):812-818.

[110] Oglesbee D, Sanders KA, Lacey JM, Magera MJ, Casetta B, Strauss KA, Tortorelli S, Rinaldo P, Matern D.

Second-tier test for quantification of alloisoleucine and branched-chain amino acids in dried blood spots to improve newborn screening for maple syrup urine disease (MSUD). Clinical chemistry. 2008;54(3):542-549.

[111] Kim JI, Noh JY, Kim M, Park JM, Song HW, Kang MJ, Pyun JC. Newborn screening by matrix-assisted laser desorption/ionization mass spectrometry based on parylene-matrix chip. Analytical biochemistry. 2017 Aug 1;530:31-39.

[112] Yunus ZM, Rahman SA, Choy YS, Keng WT, Ngu LH. Pilot study of newborn screening of inborn error of metabolism using tandem mass spectrometry in Malaysia: outcome and challenges. Journal of Pediatric Endocrinology and Metabolism. 2016;29:1031-1039. DOI: 10.1515/jpem-2016-0028

[113] Lim JS, Tan ES, John CM, Poh S, Yeo SJ, Ang JS, Adakalaisamy P, Rozalli RA, Hart C, Tan ET, Ranieri E. Inborn Error of Metabolism (IEM) screening in Singapore by electrospray ionization-tandem mass spectrometry (ESI/MS/MS): An 8 year journey from pilot to current program. Molecular genetics and metabolism. 2014;113(1-2):53-61.

[114] Filiano JJ, Bellimer SG, Kunz PL. Tandem mass spectrometry and newborn screening: pilot data and review. Pediatric Neurology 2002;26:201-204 DOI: 10.1016/s0887-8994(01)00384-8

[115] Burlina AB, Polo G, Salviati L, Duro G, Zizzo C, Dardis A, Bembi B, Cazzorla C, Rubert L, Zordan R, Desnick RJ. Newborn screening for lysosomal storage disorders by tandem mass spectrometry in North East Italy. Journal of Inherited Metabolic Disease: Official Journal of the Society for the Study of Inborn Errors of Metabolism. 2018;41:209-219. DOI:10.1007/s10545-017-0098-3

[116] GargU, Dasouki M. Expanded newborn screening of inherited metabolic disorders by tandem mass spectrometry: clinical and laboratory aspects. Clinical biochemistry. 2006;39:315-332. DOI: 10.1016/j.clinbiochem.2005.12.009

[117] Mak CM, Lee HC, Chan AY, Lam CW. Inborn errors of metabolism and expanded newborn screening: review and update. Critical reviews in clinical laboratory sciences. 2013;50:142-162. DOI: 10.3109/10408363.2013.847896

[118] Chace DH. Mass spectrometry in the clinical laboratory. Chemical reviews. 2001;101:445-478. DOI: 10.1021/cr990077+

[119] Millington DS, Kodo N, Terada N, Roe D, Chace DH. The analysis of diagnostic markers of genetic disorders in human blood and urine using tandem mass spectrometry with liquid secondary ion mass spectrometry. International Journal of Mass Spectrometry and Ion Processes. 1991;111:211-228. DOI: 10.1016/0168-1176(91)85056-R

[120] Shokri M, Karimi P, Zamanifar H, Kazemi F, Badfar G, Azami M. Phenylketonuria screening in Iranian newborns: a systematic review and meta-analysis. BMC pediatrics. 2020;20:1-6. DOI: 10.1186/s12887-020-02230-6

[121] Shibata N, Hasegawa Y, Yamada K, Kobayashi H, Purevsuren J, Yang Y, Dung VC, Khanh NN, Verma IC, Bijarnia-Mahay S, Lee DH. Diversity in the incidence and spectrum of organic acidemias, fatty acid oxidation disorders, and amino acid disorders in Asian countries: selective screening vs. expanded newborn screening. Molecular Genetics and Metabolism

Reports. 2018;16:5-10. DOI: 10.1016/j.ymgmr.2018.05.003

[122] Piras D, Locci E, Palmas F, Ferino G, Fanos V, Noto A, D'aloja E, Finco G. Rare disease: a focus on metabolomics. Expert Opinion on Orphan Drugs. 2016;4:1229-1237. DOI: 10.1080/21678707.2016.1252671

[123] Ozben T. Expanded newborn screening and confirmatory follow-up testing for inborn errors of metabolism detected by tandem mass spectrometry. Clinical Chemistry and Laboratory Medicine. 2013;51:157-176. DOI: 10.1515/cclm-2012-0472.

[124] Allard P, Grenier A, Korson MS, Zytkovicz TH. Newborn screening for hepatorenal tyrosinemia by tandem mass spectrometry: analysis of succinylacetone extracted from dried blood spots. Clinical biochemistry. 2004;37:1010-1015. DOI: 10.1016/j.clinbiochem.2004.07.006

[125] Stinton C, Geppert J, Freeman K, Clarke A, Johnson S, Fraser H, Sutcliffe P, Taylor-Phillips S. Newborn screening for Tyrosinemia type 1 using succinylacetone–a systematic review of test accuracy. Orphanet journal of rare diseases. 2017;12:1-8. DOI: 10.1186/s13023-017-0599-z

[126] Huemer, M., Kožich, V., Rinaldo, P., Baumgartner, M. R., Merinero, B., Pasquini, E., Ribes, A., & Blom, H. J. Newborn screening for homocystinurias and methylation disorders: systematic review and proposed guidelines. Journal of inherited metabolic disease, 2015;38:1007-1019. DOI: 10.1007/s10545-015-9830-z

[127] Matern D, Tortorelli S, Oglesbee D, Gavrilov D, Rinaldo P. Reduction of the false-positive rate in newborn screening by implementation of MS/MS-based second-tier tests: the Mayo Clinic experience (2004-2007). Journal of Inherited Metabolic Disease. 2007;30:585-592. DOI:10.1007/s10545-007-0691-y

[128] Hayden, Melvin R, and Suresh C Tyagi. "Homocysteine and reactive oxygen species in metabolic syndrome, type 2 diabetes mellitus, and atheroscleropathy: the pleiotropic effects of folate supplementation." Nutrition journal 2004 vol. 3 4. 10 DOI:10.1186/1475-2891-3-4

[129] Bonetti F, Brombo G, Zuliani G. The relationship between hyperhomocysteinemia and neurodegeneration. Neurodegenerative disease management. 2016;6:133-145. DOI: 10.2217/nmt-2015-0008

[130] Hua Y, Zhao H, Kong Y, Ye M. Association between the MTHFR gene and Alzheimer's disease: a meta-analysis. International Journal of Neuroscience. 2011;121:462-471. DOI:10.3109/00207454.2011.578778

[131] Yi-Le Wu, Xiu-Xiu Ding, Ye-Huan Sun, Hui-Yun Yang, Liang Sun. Methylenetetrahydrofolate reductase (MTHFR) C677T/A1298C polymorphisms and susceptibility to Parkinson's disease: A meta-analysis. Journal of the Neurological Sciences. 2013; 335:14-21 DOI: 10.1016/j.jns.2013.09.006

[132] Miletić J, Drakulić D, Pejić S, et al. Prooxidant-antioxidant balance, advanced oxidation protein products and lipid peroxidation in Serbian patients with Parkinson's disease. International Journal of Neuroscience. 2018;128:600-607. DOI:10.1080/00207454.2017.1403916

[133] J. Miletic, S. Pejic, A. Todorovic, A. Valenta Sobot, D. Drakulic, I. Pavlovic, T. Ilic, A. Stefanovic, M. Prostran, M. Stojanov. Antioxidant status and clinicopathological parameters in patients with Parkinson's disease. Vojno sanitetski pregled. 2018 DOI: 10.2298/VSP180718148M.

[134] Miletic Vukajlovic J, Drakulić D, Pejić S, Ilić TV, Stefanović A, Petković M, Schiller J. Increased plasma phosphatidylcholine/lysophos phatidylcholine ratios in patients with Parkinson's disease. Rapid Communication in Mass Spectrometry. 2019 DOI: 10.1002/rcm.8595

[135] Chace DH, Millington DS, Terada N, Kahler SG, Roe CR, Hofman LF. Rapid diagnosis of phenylketonuria by quantitative analysis for phenylalanine and tyrosine in neonatal blood spots by tandem mass spectrometry. Clinical Chemistry 1993;39:66-71.

[136] MacGloin H, Guilder L, Cleary M, Davison J, Uudelepp M, Chakrapani A. G262 Maple syrup urine disease (MSUD) referred to tertiary services before and after newborn screening (NBS). Archives of Disease in Childhood. 2019;104(Suppl 2):A106. DOI: 10.1136/archdischild-2019-rcpch.254

[137] Rashed MS, Ozand PT, Bucknall MP, Little D. Diagnosis of inborn errors of metabolism from blood spots by acylcarnitines and amino acids profiling using automated electrospray tandem mass spectrometry. Pediatric research. 1995;38:324-331. DOI: 10.1203/00006450-199509000-00009

[138] Stroek K, Boelen A, Bouva MJ, De Sain-van der Velden M, Schielen PC, Maase R, Engel H, Jakobs B, Kluijtmans LA, Mulder MF, Rubio-Gozalbo ME. Evaluation of 11 years of newborn screening for maple syrup urine disease in the Netherlands and a systematic review of the literature: Strategies for optimization. JIMD Reports. 2020 DOI: 10.1002/jmd2.12124

[139] Chace DH, Hillman SL, Millington DS, Kahler SG, Roe CR, Naylor EW. Rapid diagnosis of maple syrup urine disease in blood spots from newborns by tandem mass spectrometry. Clinical Chemistry. 1995;41:62-68. DOI: 10.1093/clinchem/41.1.62

[140] Heldt K, Schwahn B, Marquardt I, Grotzke M, Wendel U. Diagnosis of MSUD by newborn screening allows early intervention without extraneous detoxification. Molecular genetics and metabolism. 2005;84:313-316. DOI: 10.1016/j.ymgme.2004.11.010

[141] Couce ML, Ramos F, Bueno MA, Díaz J, Meavilla S, Bóveda MD, Fernandez-Marmiesse A, Garcia-Cazorla A. Evolution of maple syrup urine disease in patients diagnosed by newborn screening versus late diagnosis. European Journal of Paediatric Neurology. 2015;19:652-659. DOI: 10.1016/j.ejpn.2015.07.009

[142] Simon E, Fingerhut R, Baumkötter J, Konstantopoulou V, Ratschmann R, Wendel U. Maple syrup urine disease: favourable effect of early diagnosis by newborn screening on the neonatal course of the disease. Journal of Inherited Metabolic Disease: Official Journal of the Society for the Study of Inborn Errors of Metabolism. 2006;29:532-537. DOI: 10.1007/s10545-006-0315-y

[143] Fingerhut R, Simon E, Maier EM, Hennermann JB, Wendel U. Maple syrup urine disease: newborn screening fails to discriminate between classic and variant forms. Clinical chemistry. 2008;54:1739-1741. DOI: 10.1373/clinchem.2008.105270

[144] De Castro-Hamoy LG, Chiong MA, Estrada SC, Cordero CP. Challenges in the management of patients with maple syrup urine disease diagnosed by newborn screening in a developing country. Journal of community genetics. 2017;8:9-15. DOI: 10.1007/s12687-016-0281-5

[145] Huang HP, Chu KL, Chien YH, Wei ML, Wu ST, Wang SF, Hwu WL.

Tandem mass neonatal screening in Taiwan—report from one center. Journal of the Formosan Medical Association. 2006;105:882-886. DOI: 10.1016/S0929-6646(09)60173-X

[146] Zytkovicz TH, Fitzgerald EF, Marsden D, Larson CA, Shih VE, Johnson DM, Strauss AW, Comeau AM, Eaton RB, Grady GF. Tandem mass spectrometric analysis for amino, organic, and fatty acid disorders in newborn dried blood spots: a two-year summary from the New England Newborn Screening Program. Clinical chemistry. 2001;47:1945-1955.DOI: 10.1093/clinchem/47.11.1945

[147] Myers KA, Reeves M, Wei XC, Khan A. Cerebral edema in maple syrup urine disease despite newborn screening diagnosis and early initiation of treatment. JIMD Reports. 2012;3:103-106. DOI: 10.1007/8904_2011_69

[148] Agadi S, Sutton VR, Quach MM, Riviello Jr JJ. The electroencephalogram in neonatal maple syrup urine disease: a case report. Clinical EEG and neuroscience. 2012;43:64-67. DOI: 10.1177/1550059411429521

[149] Hassan FA, El-Mougy F, Sharaf SA, Mandour I, Morgan MF, Selim LA, Hassan SA, Salem F, Oraby A, Girgis MY, Mahmoud IG. Inborn errors of metabolism detectable by tandem mass spectrometry in Egypt: the first newborn screening pilot study. Journal of medical screening. 2016;23:124-129. DOI: 10.1177/0969141315618229

[150] Jeong JS, Sim HJ, Lee YM, Yoon HR, Kwon HJ, Hong SP. Chromatographic diagnosis of maple syrup urine disease by measuring the L-alloisoleucine/L-phenylalanine ratio in dried blood spots. Journal of Chromatography B. 2011;879:2171-2174. DOI: 10.1016/j.jchromb.2011.05.023

[151] Puckett RL, Lorey F, Rinaldo P, Lipson MH, Matern D, Sowa ME, Levine S, Chang R, Wang RY, Abdenur JE. Maple syrup urine disease: further evidence that newborn screening may fail to identify variant forms. Molecular genetics and metabolism. 2010;100:136-142. DOI: 10.1016/j.ymgme.2009.11.010

[152] Eberlin LS, Dill AL, Costa AB, Ifa DR, Cheng L, Masterson T, et al. Cholesterol Sulfate Imaging in Human Prostate Cancer Tissue by Desorption Electrospray Ionization Mass Spectrometry. Analytical Chemistry. 2010;82(9):3430-3434.

[153] Eberlin LS, Norton I, Orringer D, Dunn IF, Liu X, Ide JL, et al. Ambient mass spectrometry for the intraoperative molecular diagnosis of human brain tumors. Proceedings of the National Academy of Sciences. 2013;110(5):1611-1616.

[154] Sans M, Zhang J, Lin JQ, Feider CL, Giese N, Breen MT, et al. Performance of the MasSpec Pen for Rapid Diagnosis of Ovarian Cancer. Clinical Chemistry. 2019;65(5):674-683.

[155] Torre LA, Trabert B, DeSantis CE, Miller KD, Samimi G, Runowicz CD, et al. Ovarian cancer statistics, 2018. CA Cancer J Clin. 2018;68(4):284-296.

[156] Siegel RL, Miller KD, Jemal A. Cancer statistics, 2018. CA Cancer J Clin. 2018;68(1):7-30.

[157] Brown HM, Pirro V, Cooks RG. From DESI to the MasSpec Pen: Ambient Ionization Mass Spectrometry for Tissue Analysis and Intrasurgical Cancer Diagnosis. Clinical Chemistry. 2018;64(4):628-630.

[158] Tzafetas M, Mitra A, Paraskevaidi M, Bodai Z, Kalliala I, Bowden S, et al. The intelligent knife (iKnife) and its intraoperative diagnostic advantage for the treatment of cervical disease. Proceedings of the National Academy of Sciences. 2020;117(13):7338-7346.

[159] St John ER, Balog J, McKenzie JS, Rossi M, Covington A, Muirhead L, et al. Rapid evaporative ionisation mass spectrometry of electrosurgical vapours for the identification of breast pathology: towards an intelligent knife for breast cancer surgery. Breast Cancer Res. 2017;19(1):59.

[160] Santagata S, Eberlin LS, Norton I, Calligaris D, Feldman DR, Ide JL, Liu X, Wiley JS, Vestal ML, Ramkissoon SH, Orringer DA. Intraoperative mass spectrometry mapping of an onco-metabolite to guide brain tumor surgery. Proceedings of the National Academy of Sciences. 2014;111(30):11121-11126.

[161] Calligaris D, Norton I, Feldman DR, Ide JL, Dunn IF, Eberlin LS, Graham Cooks R, Jolesz FA, Golby AJ, Santagata S, Agar NY. Mass spectrometry imaging as a tool for surgical decision-making. Journal of Mass Spectrometry. 2013;48(11):1178-87.

[162] Basu SS, Regan MS, Randall EC, Abdelmoula WM, Clark AR, Lopez BG, Cornett DS, Haase A, Santagata S, Agar NY. Rapid MALDI mass spectrometry imaging for surgical pathology. NPJ precision oncology. 2019;3(1):1-5.

Chapter 6

Mass Spectrometry Imaging of Neurotransmitters

Katherine A. Stumpo

Abstract

Mass spectrometry imaging (MSI) is a powerful analytical method for the simultaneous analysis of hundreds of compounds within a biological sample. Despite the broad applicability of this technique, there is a critical need for advancements in methods for small molecule detection. Some molecular classes of small molecules are more difficult than others to ionize, e.g., neurotransmitters (NTs). The chemical structure of NTs (i.e., primary, secondary, and tertiary amines) affects ionization and has been a noted difficulty in the literature. In order to achieve detection of NTs using MSI, strategies must focus on either changing the chemistry of target molecules to aid in detection or focus on new methods of ionization. Additionally, even with new strategies, the issues of delocalization, chemical background noise, and ability to achieve high throughput (HTP) must be considered. This chapter will explore previous and up-and-coming techniques for maximizing the detection of NTs.

Keywords: mass spectrometry imaging, neurotransmitters, gold nanoparticles, derivatization, sample preparation

1. Introduction

Mass Spectrometry Imaging (MSI) is an incredibly powerful label-free technique that can determine qualitative and quantitative information of hundreds of compounds in a tissue section in one experiment [1, 2]. Small molecule detection, especially of neurotransmitters (NTs), currently relies heavily on histochemical, immunohistochemical, and ligand-based assays. Antibody-based methods suffer from limitations in cost and availability of antibodies, lack of specificity for target molecules of interest, and low throughput [3, 4]. The development of MSI has overcome many of these challenges and will be discussed throughout. The basic methodology of MSI is to section tissue using a cryostat to approximately 10-20 μm thickness; tissues may or may not be embedded in a cryomatrix such as Shandon™ M-1 (ThermoFisher Scientific). Next, matrix must be deposited on the tissue section, which is most often done by spray-coating the tissue using a pneumatic sprayer. Variations on typical organic matrices, such as inorganic nanoparticles (NPs), have been explored by numerous researchers and will be commented on here. Mass spectral data is collected at discrete locations on the sample surface via a raster pattern, which can then be assembled into a heat-map image of molecule location. **Figure 1** depicts the typical MSI scheme [5]. A number of overall reviews of MSI have appeared in recent years [6] that address broad topics like ionization of small molecules [7], clinical applications [2], and high-resolution analyses [5].

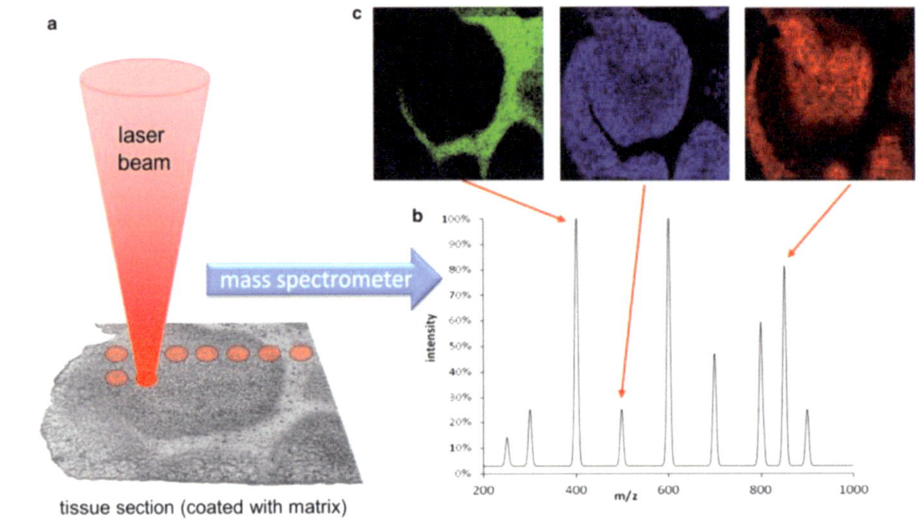

Figure 1.
General scheme of the mass spectrometry imaging process. (a) The tissue section is covered with matrix and irradiated by a pulsed laser beam. (b) Mass spectrum acquired from one spatial location on the tissue section. (c) MS images of different m/z peaks compiled from all spatial locations. Reprinted with permission from Ref. [5]. Copyright 2013 Springer.

The broader scientific community is not yet fully utilizing MSI as there are still challenges to be overcome [7, 8], including: (i) low ionization efficiency for small molecules, (ii) chemical noise interferents/overlapping signals of small molecules with traditional matrices [9], (iii) reproducibility issues across laboratories which limits universal procedures for MSI in pre-/clinical research, (iv) limits to lateral spatial resolution inherent to the matrix crystallization process which affects the ability to clearly define tissue margins, and (v) delocalization of analyte molecules during sample preparation. This chapter will focus on the analysis of small molecules, specifically neurotransmitters (NTs), due to the complex biological processes that occur in the brain and have broad implications in disease states and overall health. This chapter is broken down in two main categories, as strategies to improve ionization must either focus on (i) the chemical nature of the analyte and changing its properties to better facilitate ionization, or (ii) on utilizing a different mechanism of ionization to favor small molecules of interest.

2. Small molecule basics for MSI

Small molecule NTs are the chemical messengers of the central nervous system. Having a complete picture of NT location and abundance will aid in understanding of many different disease states and developmental processes. NTs are difficult to detect *in situ* via mass spectrometry due to their low physiological abundance (e.g., nM to pM concentration) within a complex biological tissue with many different classes of biomolecules, and overlapping low molecular mass range with most traditional matrices. Prior to the analyses discussed here, NTs were localized based on their protein-receptors or transporters, which does not always give an accurate accounting of present location.

Instrumentation used for MSI can vary widely, but most laser-based work is performed by time-of-flight (TOF) instruments. The other common setup is using desorption electrospray ionization (DESI) as an ion source, which is not the focus here, but is another option gaining in popularity [10, 11]. Other hardware configurations can favor small molecules (e.g., ion mobility, triple quadrupole instruments, Fourier transform – ion cyclotron resonance) and so there is no one-size-fits-all set-up for small molecule MSI experiments. In contrast to instrument choice, sample preparation/derivatization and ionization conditions are areas that can be standardized in order to achieve similar results across different platforms. The focus here is not on the many instrument combinations as other reviews have adequately explored this topic [6, 7].

Sample handling and preparation of tissue sections are integral to maintaining sample integrity; after cryo-sectioning, tissue is typically thaw mounted onto a solid surface. The surface must be conductive in order to apply a potential to the sample and accelerate ions out of the instrument source. Common materials include coated metal targets (expensive, cannot be archived, and not histology compatible) or indium-tin oxide (ITO) coated glass slides. Matrix application to the tissue section is ideally a homogenous coating of small crystals that provide optimum extraction conditions of analyte. After application, the key process is the co-crystallization that must occur between matrix and analyte. Spraying parameters affect the "wetness" of the surface of the tissue and are a balance between molecular diffusion and effective extraction. Crystal size is one of the more critical factors for a successful MSI experiment and multiple studies determined the parameters important for optimum crystallization.

Commonly used organic acid matrices for MSI are shown in **Table 1** and include sinapinic acid (SA), 2,5-dihydroxybenzoic acid (DHB), and α-cyano-4-hydroxycinnamic acid (CHCA). These matrices work for a broad variety of biomolecules including peptides and lipids, but do not always translate well to small molecule detection. The introduction of N-(1-napthyl)ethylenediamine dihydrochloride (NEDC) and 1,5-diaminonapthalene (DAN) have improved the detection of small molecules, though these matrices have different preparation needs (e.g., sublimation and recrystallization) which increases the time required for sample processing [12, 19]. DAN and NEDC matrices have not been fully explored yet in the literature. Moreover, the propensity for organic matrices to self-ionize and create chemical noise in the low mass range prevents effective analysis of most metabolites. Recently, this has led toward the rational design [35] or selection [36] of matrices that can address this, but the lack of consistency in performance can still be an issue. Matrix applications are notorious for behaving differently across laboratories, and significant research in the past 25 years has been devoted to identifying preparation methods that result in the most consistent data [37–39].

Sublimation procedures, mentioned *vida supra*, require that matrix and sample are placed in a vacuum chamber which is evacuated [32]. The sample is cooled while the matrix is heated, resulting in sublimation of the matrix which will condense on the cool sample surface. Recrystallization of the matrix is often coupled with this technique. There are advantages and disadvantages to all of the aforementioned matrix application techniques, which have been discussed thoroughly in the literature [40–42]. Automated sprayers have become increasingly popular and help with consistency of matrix application, though the size and spacing of the matrix droplets will ultimately affect spatial resolution of the experiment. This topic has been frequently discussed and reviewed in the literature [2, 6], so only a basic introduction is given here.

Despite these challenges, there are a few examples of successful metabolomics imaging experiments, though they have utilized purposefully designed matrices that do not generate interfering signals [35] or have used high-resolution instruments that have the high mass accuracy to distinguish between isobaric signals [43, 44]. These approaches are not an all-encompassing solution, and the next sections explore other strategies to achieve broader success with MSI of small molecules, specifically NTs.

Matrix and structure	Common analytes
Sinapinic acid (SA) 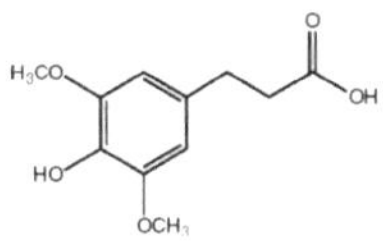	Proteins [12–15]
2,5-dihydroxybenzoic acid (DHB) 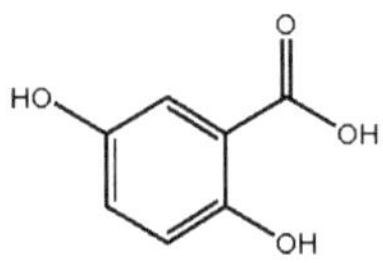	Peptides [16–18], lipids [19, 20], some small molecules [4, 21, 22]
α-cyano-4-hydroxycinnamic acid (CHCA) 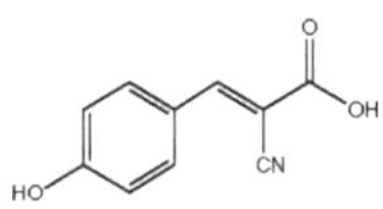	Neuropeptides [23], peptides [24], proteins [25], lipids [26, 27], drugs [28]
N-(1-napthyl)ethylenediamine dihydrochloride (NEDC)• 2HCl 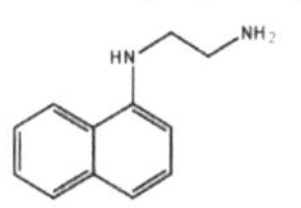	Small molecule metabolites [29], lipids [30]
1,5-diaminonapthalene (DAN) 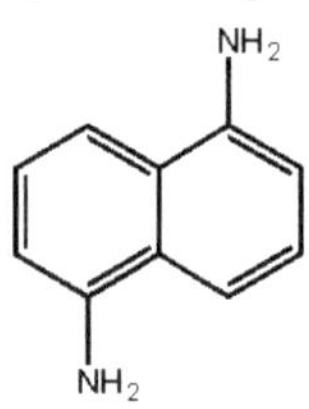	Lipids [31–33], small molecule metabolites [34]

Table 1.
Common MSI organic acid matrices and their applications.

3. Alternative matrix materials

As an alternative to the traditional organic acid matrices, contemporary studies have returned to the inorganic materials that were originally proposed for MALDI-MS by Tanaka *et al.* [45]. Nanoparticles (NPs) made of gold [46–49], silver [50–53], carbon based substrates [54–57], and silicon surfaces [58, 59] have been demonstrated on the target plate as materials that facilitate ionization of biomolecules. In particular, gold NPs (AuNPs) have the potential to be a more universal material to help facilitate ionization of small molecules and seem to have fewer reproducibility issues across multiple instrument platforms, locations, and organisms/biofluids [46, 60, 61]. Specific advantages for ionization of small molecules using AuNPs on the target plate include: (i) less chemical noise in the range where small molecules are found (below *m/z* 300), (ii) flexible analyte solution preparation conditions, including tolerances for salts, surfactants, and pH, and (iii) broad applicability across chemical classes [46, 48, 57, 60, 62].

The general success, though not broad usage, of these materials on the target plate have led to several different approaches for using alternative inorganic materials for MSI, such as sputtering of metals or the use of metallic NPs, both of which are described in the paragraphs below.

Sputtering of metals over tissue sections has produced a number of quality articles that have utilized silver [50], platinum [63], and gold [64]. Sputtering deposits highly pure and homogeneous metal or metal oxide nanolayers onto biological tissue sections. Magnetron sputtering systems utilize a plasma gun under high- or ultra-high vacuum and deposits layers of metal onto the substrate of interest. Deposition times range from under a minute to several minutes, with total sample preparation time at least several minutes long because of the need for a vacuum-based system. Sputtered layers of silver or gold are typically reported in the 20-50 nm range [49, 50, 65], which is a narrower size distribution than solution-based NPs. The biggest disadvantages of sputtering are the need for expert users, the time involved for sample preparation, and the equipment cost (e.g., sputtering apparatuses are up to tenfold more expensive than pneumatic sprayers).

Molecules that have been successfully detected using Ag or Au sputtering experiments are largely neutral lipids, with cholesterol being of high interest [49, 50, 52, 66]. Pt sputtering has been demonstrated on lipids in tissue [67] and in leaves where metabolites of interest were detected, including many with molecular features similar to NTs, such as acetamiprid [63]. Rafols *et al.* showed an Au sputtering MSI experiment that resulted in the potential detection of 25 different compounds, but only 1 of which could be called a small molecule metabolite (i.e., citrulline) [64]. A significant advantage that sputtering demonstrates, compared to organic acid matrices [22], is the lack of analyte delocalization [64]. The largest survey of sputtering materials was done by Hansen *et al.*, where Ag, Au, Cu, Ni, Pt, and Ti were sputtered for varying times on plant tissues [68]. Noble metals (e.g., Ag, Au, Pt) were found to be more effective than transition metals (e.g., Cu, Ni, Ti) for overall ionization in positive- and negative-ion modes. While lipids were the most prevalent biomolecule class examined, this is a rare demonstration of the detection of amine-based structures, including choline, asparagine, glutamic acid, and leucine. DHB was used for comparison in positive-ion mode and DAN in negative-ion mode, with primarily insoluble lipids being effectively ionized with organic matrices. A summary of the molecules detected is shown in **Figure 2**.

Nanomaterials in suspension form, such as colloidal NPs, could potentially be deposited or sprayed onto tissue sections for analysis, yet there are only select demonstrations of this application for MSI, which are described herein. This area of research has again been applied primarily to lipidomics, with successful detection of fatty acids and their derivatives, sterols, phospholipids (e.g., phosphatidylethanolamine, phosphatidylcholine, phosphatidylinositol, etc.), sphingomyelins, and ceramides. Silver NPs (AgNPs), including those that have been surface-derivatized are most popular. Polyvinylpyrrolidone-capped AgNPs have been utilized for analysis of brain [69], and AuNPs with alkylamine surface modifications have been used for imaging of glycosphingolipids in the brain [70]. Fluorinated AuNPs have been shown to detect carbohydrates, lipids, bile acids, sulfur metabolites, amino acids, nucleotide precursors, and more in mouse colon [71].

Small molecule examples, specifically NTs, are notably lacking in the MSI literature. We have been developing a method for the detection of endogenous NTs from biological samples using citrate-capped AuNPs that are pneumatically sprayed on tissue sections [47]. Successful detection of acetylcholine, dopamine/octopamine, epinephrine, glutamine, GABA, norepinephrine, and serotonin was achieved in rabbit brain tissue sections, zebrafish embryos, and neuroblastoma cells [61, 72]. See **Figure 3** for an image of seven different NTs in 5 day-post fertilization zebrafish embryos.

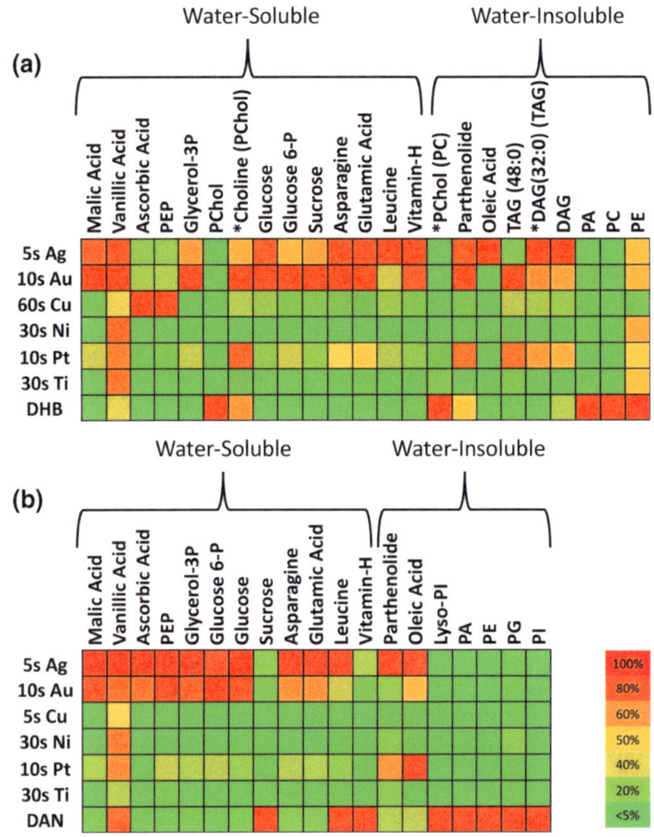

Figure 2.
Summary of sputter-coated metal screening for small metabolite analysis in (a) positive and (b) negative ion mode. Asterisks indicate a fragment ion was detected. Reprinted with permission from Ref. [68]. Copyright 2018 American Society for Mass Spectrometry.

With the intention of demonstrating the flexibility of AuNPs in terms of sample preparation, spraying parameters have been explored, including matrix concentration, solvent composition, spray temperature, and linear flow rate (which determines overall spray density of material). Early data and statistical analysis by principal component analysis (PCA) suggests that: (i) AuNP concentration can be varied over 4 orders of magnitude, (ii) a variety of organic:aqueous solvent mixes are possible, and (iii) temperatures from 30 to 60°C can be utilized [73, 74]. The flexibility in spray parameters means that less-experienced users with varying spraying capabilities can still get optimum data from their tissue sections, opening the application of MSI to more areas of study.

As previously mentioned, delocalization is an issue with organic acid matrices, often because of a "wet" matrix spray that results in true molecular diffusion instead of analyte extraction from the tissue. A standard literature method for CHCA has been compared with AuNPs, where CHCA gives only extreme delocalization outside of the tissue margins, and the AuNPs result in distinct anatomical visualization, as well as the ability to see subtle differences in analyte concentration [61]. Another advantage of AuNPs that we have discovered is flexibility in sample storage. For example, AuNP-sprayed tissue sections on slides were stored overnight at −20°C and imaging runs were repeated after 24 hours. Nearly identical data resulted and up to 8 imaging runs were completed on the same tissue section without loss of signal or the need to reapply AuNPs. The ability to archive slides for later

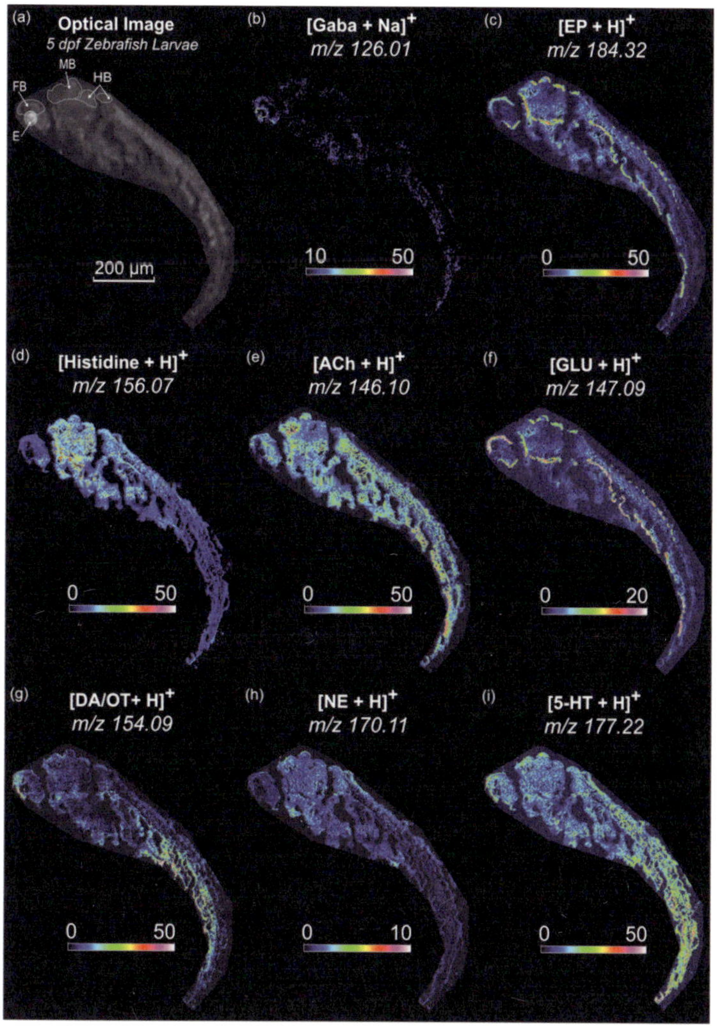

Figure 3.
MSI of a sagittal zebrafish tissue section at 5 μm lateral spatial resolution with eye (E), forebrain (FB), midbrain (MB), and hindbrain (HB) indicated in (a), (b) is the [GABA + Na]$^+$ image, (c) is the epinephrine image, (d) is the histidine image, (e) is the acetylcholine image, (f) is the glutamine image, (g) is the dopamine/octopamine image, (h) is the norepinephrine image, and (i) is the serotonin image. Reprinted with permission from Ref. [61]. Copyright 2020 American Chemical Society.

examination could allow for follow-up data to be collected as well as the potential to reduce the number of organisms needed in a given research project.

The improvements in delocalization, reproducibility, and long-term stability from pneumatically-sprayed AuNPs warrant further investigation of this method. Finally, the quick and low-cost preparation may enable a broad range of new applications in neuroscience, pharmacology, drug discovery, and pathology.

4. Derivatization strategies

Chemical derivatization of functional groups is a common strategy to improve detection in MS techniques for a variety of molecular classes. On-tissue derivatization has been explored for many different purposes, including tryptic digestion [75, 76], phospholipid digestion [77], N-terminal peptide derivatization [78], and derivatization of various metabolites/drugs of interest [79–82]. One particularly attractive advantage of derivatization is that small mass species typically have a change in molecular weight which avoids interferences from low mass matrix peaks.

Some of the aforementioned studies were on small molecular weight species, but derivatization was typically for only one analyte of interest. Examples of NT derivatization have met with varied success in terms of how many different molecular classes are accessible. Coniferyl aldehyde has been used to derivatize primary amines in pig adrenal glands and rat brains [83]. Methods included pre-coating target plates and then incubating after tissue was affixed for several minutes. Spray-coating with an organic matrix followed.

Specific reactions focused on NTs have utilized pyrylium salts (e.g., 2,4-diphenyl pyranylium (DPP)) that are reactive toward primary amines. The reaction scheme with a common NT, dopamine, is shown in **Figure 4a**. The reaction can proceed at room temperature but requires 30-80 spray passes of the derivatizing agent, followed by drying time, and then application of an organic matrix [84]. Additionally, preparation of the derivatizing agent is required and can be a multi-day process. While smaller crystals than typical organic matrix preparations have been reported, there are still limitations on overall spatial resolution. **Figure 4b** shows dopamine derivatized with 3 different pyrylium salts and the resulting images that are generated; without derivatization no dopamine was observed.

Derivatization with DPP has been demonstrated in multiple instances, with the generation a 3D mouse brain atlas of dopamine, norepinephrine and serotonin [85] as well as detection of up to 23 amino metabolites [86].

Additional derivatization methods have been developed since the initial report on primary amines only. For example, fluoromethylpyridinium-based materials

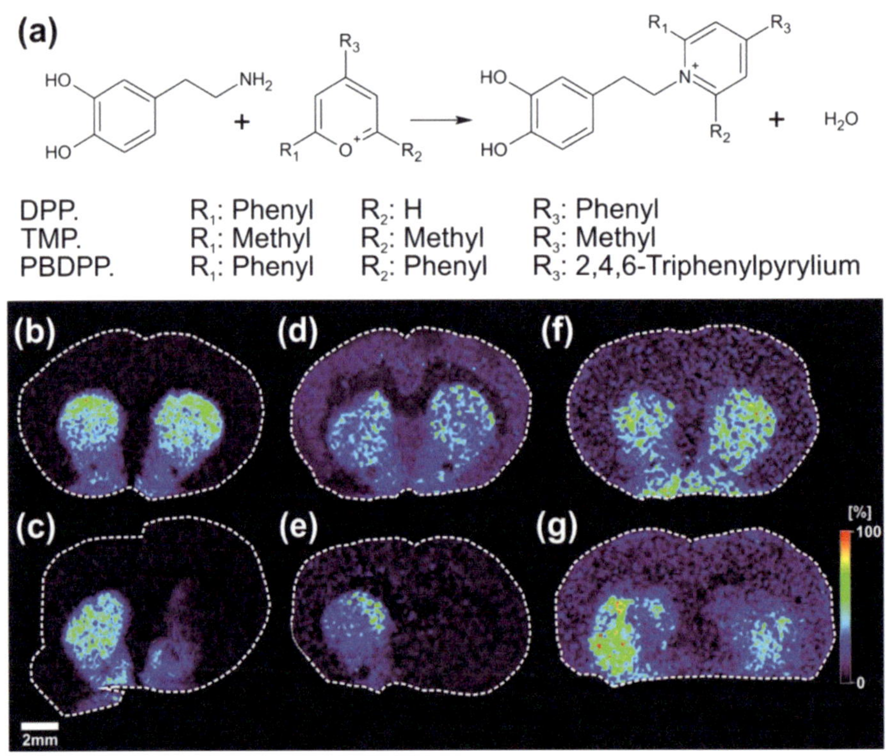

Figure 4.
(a) Reaction of dopamine with pyrylium salts. MALDI-MSI images of dopamine derivatized with DPP (b, c), PBDPP (d, e) or TMP (f, g). Signal intensity is indicated using a rainbow scale. Reprinted with permission from Ref. [84]. Copyright 2015 American Society for Mass Spectrometry.

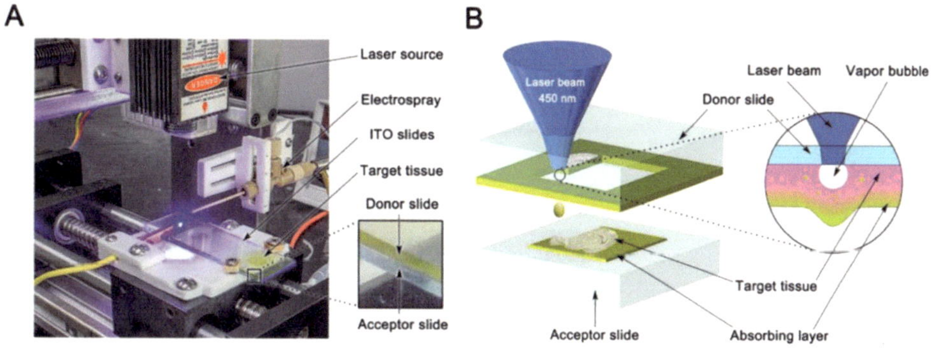

Figure 5.
(A) LATT setup and (B) schematic diagram of the system. Reprinted with permission from Ref. [89]. Copyright 2020 American Chemical Society.

are reactive with phenolic hydroxyl and/or primary or secondary amines, which expand the potential range of NTs that can be detected [87]. Charge-tagging using 2-(4-bromophenyl)-4,6- diphenylpyranylium (Br-TPP) results in distinctive isotopic distributions in the mass spectrum, making it easier to identify derivatized species from other potential species [88].

The last example here is a laser-induced tissue transfer (LATT) system that enhances on-tissue derivatization of small molecules [89]. An electrosprayer applies the derivatization reagent and matrix solution on tissue and is then irradiated with a 450 nm laser beam in transmission mode, which results in transfer of a thin film of tissue to a second slide. **Figure 5** shows the setup and diagram of the LATT system. Chemicals used for derivatization include coniferyl acetate or Girard's T reagent. Preparation time requires multiple hours (overnight) and additional matrix application. Multiple classes of biomolecules were analyzed, including amino acids, NTs, polyamines, dipeptides, and others. The issue of analyte delocalization is improved in LATT as compared to other derivatization techniques.

5. Quantitation

MSI has been applied to quantitative analysis of drugs [90, 91], metabolites [92], and biomarkers in tissue [93] using pneumatic sprayers and sublimation techniques described in this chapter. Nearly all of these demonstrations have utilized organic acid matrices such as DHB, CHCA, and trihydroxyacetophenone (THAP), with one research group utilizing TiO_2 NPs [94, 95]. Methods of quantitation are still being investigated [96], as many of the consistency issues with MALDI-MSI that have been discussed in this chapter are even more relevant with quantitative MSI (qMSI). **Figure 6** shows a summary of two of the more common methods used for generation of a calibration curve for qMSI, on-tissue spotting and tissue mimetic models which feature spiking of tissue homogenates [96].

On-tissue spotting uses either a standard molecule that is chemically similar to the analyte or a stable isotope of the analyte for making the calibration curve. Ion intensities between the analyte and standard are used to estimate the drug concentration in dosed tissue. Disadvantages include difficulty in maintaining uniform application of standards and differences in ionization for sprayed on standards vs. analyte molecules embedded within tissue. Advantages are that this method is fast and straightforward. The tissue mimetic model uses a surrogate tissue that is homogenized and spiked with the analyte of interest, frozen, sectioned, then prepared with matrix. The advantage of this method is that there is

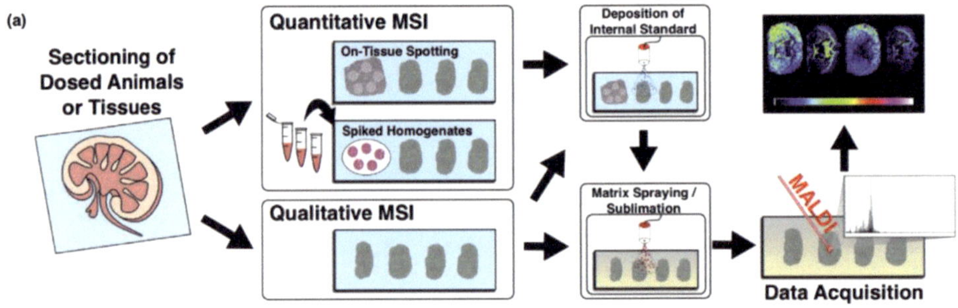

Figure 6.
Description of qMSI experiments where on-tissue spotting and homogenate spiking are two popular methods. Reprinted with permission from Ref. [94]. Copyright 2019 Elsevier Ltd.

better representation of the ionization process for analyte embedded within tissue. However, it is more time consuming, labor intensive, and requires more tissue for the calibration curve. Each method has been correlated with LC-MS data, the current primary method used for quantification [97].

Specific examples with clinical relevancy are briefly described here. First, epertinib and lapatinib were quantified in a metastatic brain cancer mouse model using stable isotope labeling, and with liquid chromatography (LC)-MS validation [91]. The topical drugs roflumilast, tofacitinib, ruxolitinib, and LEO 29102 were examined in human skin explants to determine drug penetration and evaluate lipid markers [90]. qMSI data had a much lower quantitation range than LC-MS data of individual skin layers. Rifampicin in mouse liver tissue was quantified via a fragment ion of the intact molecule. The method used an in-house synthesized stable isotope and correlated the results with LC-MS/MS [98]. Lastly, there is one example that specifically focused on comprehensive mapping of NTs in Parkinson's disease lesioned mouse brain and demonstrated quantitation of dopamine using a stable isotope [87]. All of the drug molecules listed in this paragraph are above the general size range that NTs and metabolites fall within, ranging from 400 to 800 Da, but present possible future avenues of research for the NT-focused methods discussed in this chapter.

The tissue mimetic model first gained popularity with examination of lapatinib and nevirapine in mouse liver by Groseclose and Castellino [99]. In addition to demonstrating high spatial resolution, they examined reproducibility and drug distribution within the homogenate. Fewer applications of the tissue mimetic model have been done, especially with small molecules as opposed to lipids [100]. A notable example includes the determination of the spatial distribution of gemcitabine, a chemotherapeutic agent, and its metabolites in mouse model pancreatic tumors using AuNPs and a traditional matrix as comparison [101]. Further experiments also work on the computational side of MSI and determining the best ways to normalize spectra [102, 103].

6. Concluding remarks

This chapter has introduced the utility of mass spectrometry imaging (MSI) for small molecules, with a specific focus on neurotransmitters (NTs). Methods that have resulted in enhanced signals of NTs were highlighted, with alternative matrix materials and chemical derivatization of analytes the two main points of discussion. Future research is needed in both of these areas to determine optimum conditions and applications, as well as establishing standard procedures so that broad application of MSI can continue. Finally, an area not discussed here that is relevant to these techniques and that will likely be explored in the future is the quantitative determination of small molecules.

Acknowledgements

KS thanks the University of Scranton and the Chemistry Department for providing facilities, equipment, and financial support for any referenced experiments. KS thanks the Johns Hopkins Applied Imaging Mass Spectrometry (AIMS) Core Facility at the Johns Hopkins University School of Medicine for undertaking the imaging experiments referenced in this chapter. Lastly, KS thanks Nolan McLaughlin and Tyler Bielinski for tagging along for the science journey.

Conflict of interest

The author declares no conflict of interest.

Nomenclature

DAG	diacylglycerol
DAN	1,5-diaminonapthalene
DESI	desorption electrospray ionization
DHB	2,5-dihydroxybenzoic acid
DPP	2,4-diphenyl pyranylium
CHCA	α-cyano-4-hydroxy cinnamic acid
Glucose 6-P	glαucose 6-phosphate
HTP	high throughput
ITO	indium tin oxide
LATT	laser-assisted tissue transfer
LC-MS	liquid chromatography mass spectrometry
MALDI	matrix-assisted laser desorption ionization
MSI	mass spectrometry imaging
qMSI	quantitative mass spectrometry imaging
NEDC	N-(1-napthyl)ethylenediamine dihydrochloride
NT	neurotransmitter
NP	nanoparticle
PA	phosphatidic acid
PCA	principal component analysis
PChol	phosphocholine
PE	phospatidylethanolamine
PEP	phosphoenolpyruvic acid
PG	phosphatidylglycerol
PI	phosphatidylinositol
PC	phosphatidylcholine
SA	sinapic (or sinapinic) acid
TAG	triacylglycerol
THAP	trihydroxyacetophenone

Author details

Katherine A. Stumpo
University of Scranton, Scranton, PA, USA

*Address all correspondence to: katherine.stumpo@scranton.edu

References

[1] Caprioli, R.M., Farmer, T.B., Gile, J.: Molecular Imaging of Biological Samples: Localization of Peptides and Proteins Using MALDI-TOF MS. Anal. Chem. 69, 4751-4760 (1997). https://doi.org/10.1021/ac970888i

[2] Norris, J.L., Caprioli, R.M.: Analysis of tissue specimens by matrix-assisted laser desorption/ionization imaging mass spectrometry in biological and clinical research. Chem. Rev. 113, 2309-2342 (2013). https://doi.org/10.1021/cr3004295

[3] Danbolt, N.C., Zhou, Y., Furness, D.N., Holmseth, S.: Strategies for immunohistochemical protein localization using antibodies: What did we learn from neurotransmitter transporters in glial cells and neurons. Glia. 64, 2045-2064 (2016). https://doi.org/10.1002/glia.23027

[4] Ye, H., Wang, J., Greer, T., Strupat, K., Li, L.: Visualizing Neurotransmitters and Metabolites in the Central Nervous System by High Resolution and High Accuracy Mass Spectrometric Imaging. ACS Chem. Neurosci. 4, 1049-1056 (2013). https://doi.org/10.1021/cn400065k

[5] Römpp, A., Spengler, B.: Mass spectrometry imaging with high resolution in mass and space, (2013)

[6] Buchberger, A.R., DeLaney, K., Johnson, J., Li, L.: Mass Spectrometry Imaging: A Review of Emerging Advancements and Future Insights, (2018)

[7] Trim, P.J., Snel, M.F.: Small molecule MALDI MS imaging: Current technologies and future challenges. Methods. 104, 127-141 (2016). https://doi.org/10.1016/j.ymeth.2016.01.011

[8] Seeley, E.H., Caprioli, R.M.: MALDI imaging mass spectrometry of human tissue: method challenges and clinical perspectives. Trends Biotechnol. 29, 136-143 (2011). https://doi.org/10.1016/j.tibtech.2010.12.002

[9] Smirnov, I.P., Zhu, X., Taylor, T., Huang, Y., Ross, P., Papayanopoulos, I.A., Martin, S.A., Pappin, D.J.: Suppression of α-Cyano-4-hydroxycinnamic Acid Matrix Clusters and Reduction of Chemical Noise in MALDI-TOF Mass Spectrometry. Anal. Chem. 76, 2958-2965 (2004). https://doi.org/10.1021/ac035331j

[10] Manicke, N.E., Dill, A.L., Ifa, D.R., Cooks, R.G.: High-resolution tissue imaging on an orbitrap mass spectrometer by desorption electrospray ionization mass spectrometry. J. Mass Spectrom. 45, 223-226 (2010). https://doi.org/10.1002/jms.1707

[11] Wiseman, J., Ifa, D.R., Song, Q., Cooks, R.G.: Tissue Imaging at Atmospheric Pressure Using Desorption Electrospray Ionization (DESI) Mass Spectrometry. Angew. Chemie Int. Ed. 45, 7188-7192 (20016). https://doi.org/10.1002/anie.200602449

[12] Yang, J., Caprioli, R.M.: Matrix sublimation/recrystallization for imaging proteins by mass spectrometry at high spatial resolution. Anal. Chem. 83, 5728-5734 (2011). https://doi.org/10.1021/ac200998a

[13] Leinweber, B.D., Tsaprailis, G., Monks, T.J., Lau, S.S.: Improved MALDI-TOF Imaging Yields Increased Protein Signals at High Molecular Mass. J. Am. Soc. Mass Spectrom. 20, 89-95 (2009). https://doi.org/10.1016/j.jasms.2008.09.008

[14] Mainini, V., Bovo, G., Chinello, C., Gianazza, E., Grasso, M., Cattoretti, G., Magni, F.: Detection of high molecular weight proteins by MALDI imaging mass spectrometry. Mol. Biosyst. 9, 1101-1107 (2013). https://doi.org/10.1039/c2mb25296a

[15] Grey, A.C., Chaurand, P., Caprioli, R.M., Schey, K.L.: MALDI imaging mass spectrometry of integral membrane proteins from ocular lens and retinal tissue. J. Proteome Res. 8, 3278-3283 (2009). https://doi.org/10.1021/pr800956y

[16] Andersson, M., Groseclose, M.R., Deutch, A.Y., Caprioli, R.M.: Imaging mass spectrometry of proteins and peptides: 3D volume reconstruction. Nat. Methods. 5, 101-108 (2008). https://doi.org/10.1038/nmeth1145

[17] Groseclose, M.R., Andersson, M., Hardesty, W.M., Caprioli, R.M.: Identification of proteins directly from tissue: In situ tryptic digestions coupled with imaging mass spectrometry. J. Mass Spectrom. 42, 254-262 (2007). https://doi.org/10.1002/jms.1177

[18] Altelaar, A.F.M., Taban, I.M., McDonnell, L.A., Verhaert, P.D.E.M., de Lange, R.P.J., Adan, R.A.H., Mooi, W.J., Heeren, R.M.A., Piersma, S.R.: High-resolution MALDI imaging mass spectrometry allows localization of peptide distributions at cellular length scales in pituitary tissue sections. Int. J. Mass Spectrom. 260, 203-211 (2007). https://doi.org/10.1016/j.ijms.2006.09.028

[19] Dufresne, M., Patterson, N.H., Norris, J.L., Caprioli, R.M.: Combining Salt Doping and Matrix Sublimation for High Spatial Resolution MALDI Imaging Mass Spectrometry of Neutral Lipids. Anal. Chem. 91, 12928-12934 (2019). https://doi.org/10.1021/acs.analchem.9b02974

[20] Schiller, J., Arnhold, J., Benard, S., Müller, M., Reichl, S., Arnold, K.: Lipid analysis by matrix-assisted laser desorption and ionization mass spectrometry: A methodological approach. Anal. Biochem. 267, 46-56 (1999). https://doi.org/10.1006/abio.1998.3001

[21] Dekker, L.J.M., van Kampen, J.J.A., Reedijk, M.L., Burgers, P.C., Gruters, R.A., Osterhaus, A.D.M.E., Luider, T.M.: A mass spectrometry based imaging method developed for the intracellular detection of HIV protease inhibitors. Rapid Commun. Mass Spectrom. 23, 1183-1188 (2009). https://doi.org/10.1002/rcm.3981

[22] Li, B., Zhang, Y., Ge, J., Liu, K., Li, P.: Sample preparation for mass spectrometry imaging of leaf tissues: a case study on analyte delocalization. Anal Bioanal Chem. 410, 7449-7456 (2018)

[23] Ly, A., Ragionieri, L., Liessem, S., Becker, M., Deininger, S.O., Neupert, S., Predel, R.: Enhanced Coverage of Insect Neuropeptides in Tissue Sections by an Optimized Mass-Spectrometry-Imaging Protocol. Anal. Chem. 91, 1980-1988 (2019). https://doi.org/10.1021/acs.analchem.8b04304

[24] Gemperline, E., Keller, C., Jayaraman, D., Maeda, J., Sussman, M.R., Ané, J.M., Li, L.: Examination of Endogenous Peptides in *Medicago truncatula* Using Mass Spectrometry Imaging. J. Proteome Res. 15, 4403-4411 (2016). https://doi.org/10.1021/acs.jproteome.6b00471

[25] Goodwin, R.J.A., Pennington, S.R., Pitt, A.R.: Protein and peptides in pictures: Imaging with MALDI mass spectrometry. Proteomics. 8, 3785-3800 (2008). https://doi.org/10.1002/pmic.200800320

[26] Hart, P.J., Francese, S., Claude, E., Woodroofe, M.N., Clench, M.R.: MALDI-MS imaging of lipids in ex vivo human skin. Anal. Bioanal. Chem. 401, 115-125 (2011). https://doi.org/10.1007/s00216-011-5090-4

[27] Kompauer, M., Heiles, S., Spengler, B.: Atmospheric pressure MALDI mass spectrometry imaging of tissues and cells at 1.4-μm lateral resolution. Nat. Methods. 14, 90-96 (2016). https://doi.org/10.1038/nmeth.4071

[28] Marko-Varga, G., Fehniger, T.E., Rezeli, M., Döme, B., Laurell, T., Végvári, Á.: Drug localization in different lung cancer phenotypes by MALDI mass spectrometry imaging. J. Proteomics. 74, 982-992 (2011). https://doi.org/10.1016/j.jprot.2011.03.019

[29] Wang, J., Qiu, S., Chen, S., Xiong, C., Liu, H., Wang, J., Zhang, N., Hou, J., He, Q., Nie, Z.: MALDI-TOF MS imaging of metabolites with a N -(1-Naphthyl) ethylenediamine dihydrochloride matrix and its application to colorectal cancer liver metastasis. Anal. Chem. 87, 422-430 (2015). https://doi.org/10.1021/ac504294s

[30] Wang, J., Wang, C., Han, X.: Enhanced coverage of lipid analysis and imaging by matrix-assisted laser desorption/ionization mass spectrometry via a strategy with an optimized mixture of matrices. Anal. Chim. Acta. 1000, 155-162 (2018). https://doi.org/10.1016/j.aca.2017.09.046

[31] Thomas, A., Charbonneau, J.L., Fournaise, E., Chaurand, P.: Sublimation of new matrix candidates for high spatial resolution imaging mass spectrometry of lipids: Enhanced information in both positive and negative polarities after 1,5-diaminonapthalene deposition. Anal. Chem. 84, 2048-2054 (2012). https://doi.org/10.1021/ac2033547

[32] Fernández, R., Garate, J., Martín-Saiz, L., Galetich, I., Fernández, J.A.: Matrix Sublimation Device for MALDI Mass Spectrometry Imaging. Anal. Chem. 91, 803-807 (2019). https://doi.org/10.1021/acs.analchem.8b04765

[33] Kaya, I., Jennische, E., Dunevall, J., Lange, S., Ewing, A.G., Malmberg, P., Baykal, A.T., Fletcher, J.S.: Spatial Lipidomics Reveals Region and Long Chain Base Specific Accumulations of Monosialogangliosides in Amyloid Plaques in Familial Alzheimer's Disease Mice (5xFAD) Brain. ACS Chem. Neurosci. 11, 14-24 (2020). https://doi.org/10.1021/acschemneuro.9b00532

[34] Korte, A.R., Lee, Y.J.: MALDI-MS analysis and imaging of small molecule metabolites with 1,5-diaminonaphthalene (DAN). J. Mass Spectrom. 49, 737-741 (2014). https://doi.org/10.1002/jms.3400

[35] Weißflog, J., Svatoš, A.: 1,8-Di(piperidinyl)-naphthalene-rationally designed MAILD/MALDI matrix for metabolomics and imaging mass spectrometry. RSC Adv. 6, 75073-75081 (2016). https://doi.org/10.1039/c6ra17237g

[36] Demeure, K., Quinton, L., Gabelica, V., De Pauw, E.: Rational Selection of the Optimum MALDI Matrix for Top-Down Proteomics by In-Source Decay. Anal. Chem. 79, 8678-8685 (2007). https://doi.org/10.1021/ac070849z

[37] Gusev, A.I., Wilkinson, W.R., Proctor, A., Hercules, D.M.: Improvement of signal reproducibility and matrix/comatrix effects in MALDI analysis. Anal. Chem. 67, 1034-1041 (1995). https://doi.org/10.1021/ac00102a003

[38] O'Rourke, M.B., Djordjevic, S.P., Padula, M.P.: The quest for improved reproducibility in MALDI mass spectrometry. Mass Spectrom. Rev. 37, 217-228 (2018). https://doi.org/10.1002/mas.21515

[39] Calvano, C.D., Monopoli, A., Cataldi, T.R.I., Palmisano, F.: MALDI matrices for low molecular weight compounds: an endless story? Anal. Bioanal. Chem. 410, 4015-4038 (2018). https://doi.org/10.1007/s00216-018-1014-x

[40] Gemperline, E., Rawson, S., Li, L.: Optimization and comparison of multiple MALDI matrix application methods for small molecule mass spectrometric imaging. Anal. Chem. 86, 10030-10035 (2014). https://doi.org/10.1021/ac5028534

[41] Tucker, L.H., Conde-González, A., Cobice, D., Hamm, G.R., Goodwin, R.J.A., Campbell, C.J., Clarke, D.J., Mackay, C.L.: MALDI Matrix Application Utilizing a Modified 3D Printer for Accessible High Resolution Mass Spectrometry Imaging. Anal. Chem. 90, 8742-8749 (2018). https://doi.org/10.1021/acs.analchem.8b00670

[42] Goodwin, R.J.A.: Sample preparation for mass spectrometry imaging: Small mistakes can lead to big consequences. J. Proteomics. (2012). https://doi.org/10.1016/j.jprot.2012.04.012

[43] Kreutzer, L., Aichler, M., Walch, A.K.: In situ metabolomics in cancer tissue by high-resolution mass spectrometry imaging. In: Fundamentals and Applications of Fourier Transform Mass Spectrometry. pp. 253-279. Elsevier (2019)

[44] Ly, A., Buck, A., Balluff, B., Sun, N., Gorzolka, K., Feuchtinger, A., Janssen, K.P., Kuppen, P.J.K., Van De Velde, C.J.H., Weirich, G., Erlmeier, F., Langer, R., Aubele, M., Zitzelsberger, H., McDonnell, L., Aichler, M., Walch, A.: High-mass-resolution MALDI mass spectrometry imaging of metabolites from formalin-fixed paraffin-embedded tissue. Nat. Protoc. 11, 1428-1443 (2016). https://doi.org/10.1038/nprot.2016.081

[45] Tanaka, K., Waki, H., Ido, Y., Akita, S., Yoshida, Y., Yoshida, T., Matsuo, T.: Protein and polymer analyses up to m/z 100 000 by laser ionization time-of-flight mass spectrometry. Rapid Commun. Mass Spectrom. 2, 151-153 (1988). https://doi.org/10.1002/rcm.1290020802

[46] McLean, J.A., Stumpo, K.A., Russell, D.H.: Size-Selected (2-10 nm) Gold Nanoparticles for Matrix Assisted Laser Desorption Ionization of Peptides. J. Am. Chem. Soc. 127, 5304-5305 (2005). https://doi.org/10.1021/ja043907w

[47] Stumpo, K.A., McLaughlin, N.: Method of Comprehensive Neuro-transmitter Detection using citrate-capped AuNPs for Imaging Mass Spectrometry. Proceedings of the 67th American Society for Mass Spectrometry National Meeting (2019)

[48] Wu, H.-P., Yu, C.-J., Lin, C.-Y., Lin, Y.-H., Tseng, W.-L.: Gold nanoparticles as assisted matrices for the detection of biomolecules in a high-salt solution through laser desorption/ionization mass spectrometry. J. Am. Soc. Mass Spectrom. 20, 875-882 (2009). https://doi.org/10.1016/j.jasms.2009.01.002

[49] Dufresne, M., Masson, J.-F., Chaurand, P.: Sodium-Doped Gold-Assisted Laser Desorption Ionization for Enhanced Imaging Mass Spectrometry of Triacylglycerols from Thin Tissue Sections. Anal. Chem. 88, 6018-6025 (2016). https://doi.org/10.1021/acs.analchem.6b01141

[50] Dufresne, M., Thomas, A., Breault-Turcot, J., Masson, J.F., Chaurand, P.: Silver-assisted laser desorption ionization for high spatial resolution imaging mass spectrometry of olefins from thin tissue sections. Anal. Chem. 85, 3318-3324 (2013). https://doi.org/10.1021/ac3037415

[51] Sherrod, S.D., Diaz, A.J., Russell, W.K., Cremer, P.S., Russell, D.H.: Silver Nanoparticles as Selective Ionization Probes for Analysis of Olefins by Mass Spectrometry. Anal. Chem. 80, 6796-6799 (2008). https://doi.org/10.1021/ac800904g

[52] Muller, L., Kailas, A., Jackson, S.N., Roux, A., Barbacci, D.C., Albert Schultz, J., Balaban, C.D., Woods, A.S., Schultz, J.A., Balaban, C.D., Woods, A.S., Albert Schultz, J., Balaban, C.D., Woods, A.S.: Lipid imaging within the normal rat kidney using silver nanoparticles by matrix-assisted laser desorption/ionization mass

spectrometry. Kidney Int. 88, 186-192 (2015). https://doi.org/10.1038/ki.2015.3

[53] Hua, L., Chen, J., Ge, L., Tan, S.N.: Silver nanoparticles as matrix for laser desorption/ionization mass spectrometry of peptides. J. Nanoparticle Res. 9, 1133-1138 (2007). https://doi.org/10.1007/s11051-007-9244-4

[54] Pan, C., Xu, S., Hu, L., Su, X., Ou, J., Zou, H., Guo, Z., Zhang, Y., Guo, B.: Using Oxidized Carbon Nanotubes as Matrix for Analysis of Small Molecules by MALDI-TOF MS. J. Am. Soc. Mass Spectrom. 16, 883-892 (2005). https://doi.org/10.1016/j.jasms.2005.03.009

[55] Ren, S., Zhang, L., Cheng, Z., Guo, Y.: Immobilized carbon nanotubes as matrix for MALDI-TOF-MS analysis: Applications to neutral small carbohydrates. J. Am. Soc. Mass Spectrom. 16, 333-339 (2005). https://doi.org/10.1016/j.jasms.2004.11.017

[56] Han, M., Sunner, J.: An activated carbon substrate surface for laser desorption mass spectrometry. J. Am. Soc. Mass Spectrom. 11, 644-649 (2000). https://doi.org/10.1016/S1044-0305(00)00129-X

[57] Tang, H.-W., Ng, K.-M., Lu, W., Che, C.-M.: Ion Desorption Efficiency and Internal Energy Transfer in Carbon-Based Surface-Assisted Laser Desorption/Ionization Mass Spectrometry: Desorption Mechanism(s) and the Design of SALDI Substrates. Anal. Chem. 81, 4720-4729 (2009). https://doi.org/10.1021/ac8026367

[58] Wei, J., Buriak, J.M., Siuzdak, G.: Desorption–ionization mass spectrometry on porous silicon. Nature. 399, 243-246 (1999). https://doi.org/10.1038/20400

[59] Wen, X., Dagan, S., Wysocki, V.H.: Small-Molecule Analysis with Silicon-Nanoparticle-Assisted Laser Desorption/Ionization Mass Spectrometry. Anal. Chem. 79, 434-444 (2007). https://doi.org/10.1021/ac061154l

[60] Sacks, C., Stumpo, K.A.: Gold Nanoparticles for Enhanced Ionization and Fragmentation of Biomolecules using LDI-MS. J. Mass. Spectrom. 53, 1070-1077 (2018). https://doi.org/10.1002/jms.4282

[61] McLaughlin, N., Bielinski, T., Tressler, C., Barton, E., Glunde, K., Stumpo, K.A.: Pneumatically Sprayed Gold Nanoparticles for Mass Spectrometry Imaging of Neurotransmitters. J. Am. Soc. Mass Spectrom. ASAP Article, DOI: 10.1021/jasms.0c00156

[62] Chiu, T., Chang, L., Chiang, C., Chang, H.: Determining Estrogens Using Surface-Assisted Laser Desorption/Ionization Mass Spectrometry with Silver Nanoparticles as the Matrix. J. Am. Soc. Mass Spectrom. 19, 1343-1346 (2008). https://doi.org/10.1016/j.jasms.2008.06.006

[63] Ozawa, T., Osaka, I., Hamada, S., Murakami, T., Miyazato, A., Kawasaki, H., Arakawa, R.: Direct Imaging Mass Spectrometry of Plant Leaves Using Surface-assisted Laser Desorption/Ionization with Sputter-deposited Platinum Film. Anal. Sci. 32, 587-591 (2016). https://doi.org/10.2116/analsci.32.587

[64] Ràfols, P., Vilalta, D., Torres, S., Calavia, R., Heijs, B., McDonnell, L.A., Brezmes, J., del Castillo, E., Yanes, O., Ramírez, N., Correig, X.: Assessing the potential of sputtered gold nanolayers in mass spectrometry imaging for metabolomics applications. PLoS One. 13, (2018). https://doi.org/10.1371/journal.pone.0208908

[65] Xu, L., Kliman, M., Forsythe, J.G., Korade, Z., Hmelo, A.B., Porter, N.A., McLean, J.A.: Profiling and Imaging Ion

Mobility-Mass Spectrometry Analysis of Cholesterol and 7-Dehydrocholesterol in Cells Via Sputtered Silver MALDI. J. Am. Soc. Mass Spectrom. 26, 924-933 (2015). https://doi.org/10.1007/s13361-015-1131-0

[66] Tang, H.-W., Lu, W., Che, C.-M., Ng, K.-M.: Gold Nanoparticles and Imaging Mass Spectrometry: Double Imaging of Latent Fingerprints. Anal. Chem. 82, 1589-1593 (2010). https://doi.org/10.1021/ac9026077

[67] Ozawa, T., Osaka, I., Ihozaki, T., Hamada, S., Kuroda, Y., Murakami, T., Miyazato, A., Kawasaki, H., Arakawa, R.: Simultaneous detection of phosphatidylcholines and glycerolipids using matrix-enhanced surface-assisted laser desorption/ionization-mass spectrometry with sputter-deposited platinum film. J. Mass. Spectrom. 50, 1264-1269 (2015)

[68] Hansen, R.L., Duenas, M.E., Lee, Y.J.: Sputter-Coated Metal Screening for Small Molecule Analysis and High-Spatial Resolution Imaging in Laser Desorption Ionization Mass Spectrometry | Journal of the American Society for Mass Spectrometry. J. Am. Soc. Mass Spectrom. 30, 299-308 (2019)

[69] Guan, M., Zhang, Z., Li, S., Liu, J., Liu, L., Yang, H., Zhang, Y., Wang, T., Zhao, Z.: Silver nanoparticles as matrix for MALDI FTICR MS profiling and imaging of diverse lipids in brain. Talanta. 179, 624-631 (2018)

[70] Goto-Inoue, N., Hayasaka, T., Zaima, N., Kashiwagi, Y., Yamamoto, M., Nakamoto, M., Setou, M.: The detection of glycosphingolipids in brain tissue sections by imaging mass spectrometry using gold nanoparticles. J. Am. Soc. Mass Spectrom. 21, 1940-1943 (2010). https://doi.org/10.1016/j.jasms.2010.08.002

[71] Palermo, A., Forsberg, E.M., Warth, B., Aisporna, A.E., Billings, E., Kuang, E., Benton, H.P., Berry, D., Siuzdak, G.: Fluorinated Gold Nanoparticles for Nanostructure Imaging Mass Spectrometry. ACS Nano. 12, 6938-6948 (2018). https://doi.org/10.1021/acsnano.8b02376

[72] McLaughlin, N., Bielinski, T., Stumpo, K.A.: The new gold standard for mass spectrometry imaging. In: Proceedings of the 68th American Society for Mass Spectrometry National Meeting (2020)

[73] Bielinski, T.: Optimization and Method Development of Pneumatically Sprayed AuNPs for Laser-Desorption Ionization Mass Spectrometry Imaging. Proceedings of the 68th American Society for Mass Spectrometry National Meeting (2020)

[74] Shedlock, C., Gianello, M., Bielinski, T., McLaughlin, N., Graham, J., Stumpo, K.A.: Computational Analysis of MSI Preparation Conditions using AuNPs. Proceedings of the 68th American Society for Mass Spectrometry National Meeting. Current Opinion in Biotechnology. (2020), 51-59. DOI: 10.1016/j.copbio.2018.08.003

[75] Beine, B., Diehl, H.C., Meyer, H.E., Henkel, C.: Tissue maldi mass spectrometry imaging (MALDI MSI) of peptides. In: Methods in Molecular Biology. pp. 129-150. Humana Press Inc. (2016)

[76] Diehl, H.C., Beine, B., Elm, J., Trede, D., Ahrens, M., Eisenacher, M., Marcus, K., Meyer, H., Henkel, C.: The challenge of on-tissue digestion for MALDI MSI— a comparison of different protocols to improve imaging experiments. Anal Bioanal Chem. 407, 2223-2243 (2015)

[77] Vens-Cappell, S., U. Kouzel, I., Kettling, H., Soltwisch, J., Bauwens, A., Porubsky, S., Müthing, J., Dreisewerd, K.: On-Tissue Phospholipase C Digestion for Enhanced MALDI-MS Imaging

of Neutral Glycosphingolipids. Anal. Chem. 88, 5595-5599 (2016). https://doi.org/10.1021/acs.analchem.6b01084

[78] Franck, J., El Ayed, M., Wisztorski, M., Salzet, M., Fournier, I.: On-Tissue N-Terminal Peptide Derivatizations for Enhancing Protein Identification in MALDI Mass Spectrometric Imaging Strategies. Anal. Chem. 81, 8305-8317 (2009). https://doi.org/10.1021/ac901043n

[79] Flinders, B., Morrell, J., Marshall, P.S., Ranshaw, L.E., Clench, M.R.: The use of hydrazine-based derivatization reagents for improved sensitivity and detection of carbonyl containing compounds using MALDI-MSI. Anal Bioanal Chem. 407, 2085-2094 (2015)

[80] Fülöp, A., Bausbacher, T., Rizzo, S., Zhou, Q., Gillandt, H., Hopf, C., Rittner, M.: New Derivatization Reagent for Detection of free Thiol-groups in Metabolites and Proteins in Matrix-Assisted Laser Desorption/Ionization Mass Spectrometry Imaging. Anal. Chem. 92, 6224-6228 (2020). https://doi.org/10.1021/acs.analchem.9b05630

[81] Shimma, S., Kumada, H.-O., Taniguchi, H., Konna, A., Yao, I., Furata, K., Matsuda, T., Ito, S.: Microscopic visualization of testosterone in mouse testis by use of imaging mass spectrometry. Anal Bioanal Chem. 408, 7607-7615 (2016)

[82] P. Y. Barré, F., Flinders, B., P. Garcia, J., Jansen, I., R. S. Huizing, L., Porta, T., B. Creemers, L., M. A. Heeren, R., Cillero-Pastor, B.: Derivatization Strategies for the Detection of Triamcinolone Acetonide in Cartilage by Using Matrix-Assisted Laser Desorption/Ionization Mass Spectrometry Imaging. Anal. Chem. 88, 12051-12059 (2016). https://doi.org/10.1021/acs.analchem.6b02491

[83] Manier, M.L., Spraggins, J.M., Reyzer, M.L., Norris, J.L., Caprioli, R.M.: A derivatization and validation strategy for determining the spatial localization of endogenous amine metabolites in tissues using MALDI imaging mass spectrometry: Endogenous amine metabolites in tissue. J. Mass Spectrom. 49, 665-673 (2014). https://doi.org/10.1002/jms.3411

[84] Shariatgorji, M., Nilsson, A., Källback, P., Karlsson, O., Zhang, X., Svenningsson, P., Andren, P.E.: Pyrylium Salts as Reactive Matrices for MALDI-MS Imaging of Biologically Active Primary Amines. J. Am. Soc. Mass Spectrom. 26, 934-939 (2015). https://doi.org/10.1007/s13361-015-1119-9

[85] Sugiyama, E., Guerrini, M.M., Honda, K., Hattori, Y., Abe, M., Källback, P., Andrén, P.E., Tanaka, K.F., Setou, M., Fagarasan, S., Suematsu, M., Sugiura, Y.: Detection of a high turnover serotonin circuit in the mouse brain using mass spectrometry imaging. iScience. (2019). https://doi.org/10.1016/j.isci.2019.09.036

[86] Esteve, C., Tolner, E.A., Shyti, R., van den Maagdenberg, A.M.J.M., McDonnell, L.A.: Mass spectrometry imaging of amino neurotransmitters: a comparison of derivatization methods and application in mouse brain tissue. Metabolomics. 12, 1-9 (2016). https://doi.org/10.1007/s11306-015-0926-0

[87] Shariatgorji, M., Nilsson, A., Fridjonsdottir, E., Vallianatou, T., Källback, P., Katan, L., Sävmarker, J., Mantas, I., Zhang, X., Bezard, E., Svenningsson, P., Odell, L.R., Andrén, P.E.: Comprehensive mapping of neurotransmitter networks by MALDI–MS imaging. Nat. Methods. 16, 1021-1028 (2019). https://doi.org/10.1038/s41592-019-0551-3

[88] Shariatgorji, R., Nilsson, A., Strittmatter, N., Vallianatou, T., Zhang, X., Svenningsson, P., Goodwin, R.J.A., Andrén, P.E.: Bromopyrylium Derivatization Facilitates Identification

by Mass Spectrometry Imaging of Monoamine Neurotransmitters and Small Molecule Neuroactive Compounds. J. Am. Soc. Mass Spectrom. 16, 18 (2020). https://doi.org/10.1021/jasms.0c00166

[89] Guo, S., Tang, W., Hu, Y., Chen, Y., Gordon, A., Li, B., Li, P.: Enhancement of On-tissue Chemical Derivatization by LaserAssisted Tissue Transfer for MALDI MS Imaging. Anal. Chem. 92, 1431-1438 (2020)

[90] Bonnel, D., Legouffe, R., Eriksson, A.H., Mortensen, R.W., Pamelard, F., Stauber, J., Nielsen, K.T.: MALDI imaging facilitates new topical drug development process by determining quantitative skin distribution profiles. Anal. Bioanal. Chem. 410, 2815-2828 (2018). https://doi.org/10.1007/s00216-018-0964-3

[91] Tanaka, Y., Hirata, M., Shinonome, S., Torii, M., Nezasa, K.I., Tanaka, H.: Distribution analysis of epertinib in brain metastasis of HER2-positive breast cancer by imaging mass spectrometry and prospect for antitumor activity. Sci. Rep. 8, 1-12 (2018). https://doi.org/10.1038/s41598-017-18702-2

[92] Goodwin, R.J.A., Nilsson, A., Mackay, C.L., Swales, J.G., Johansson, M.K., Billger, M., Andrén, P.E., Iverson, S.L.: Exemplifying the Screening Power of Mass Spectrometry Imaging over Label-Based Technologies for Simultaneous Monitoring of Drug and Metabolite Distributions in Tissue Sections. J. Biomol. Screen. 21, 187-193 (2016). https://doi.org/10.1177/1087057115623740

[93] Hulme, H., Fridjonsdottir, E., Gunnarsdottir, H., Vallianatou, T., Zhang, X., Wadensten, H., Shariatgorji, R., Nilsson, A., Bezard, E., Svenningsson, P., Andrén, P.E.: Simultaneous mass spectrometry imaging of multiple neuropeptides in the brain and alterations induced by experimental parkinsonism and L-DOPA therapy. Neurobiol. Dis. 137, 104738 (2020). https://doi.org/10.1016/j.nbd.2020.104738

[94] Giordano, S., Zucchetti, M., Decio, A., Cesca, M., Fuso Nerini, I., Maiezza, M., Ferrari, M., Licandro, S.A., Frapolli, R., Giavazzi, R., Maurizio, D., Davoli, E., Morosi, L.: Heterogeneity of paclitaxel distribution in different tumor models assessed by MALDI mass spectrometry imaging. Sci. Rep. 6, 1-12 (2016). https://doi.org/10.1038/srep39284

[95] Giordano, S., Morosi, L., Veglianese, P., Licandro, S.A., Frapolli, R., Zucchetti, M., Cappelletti, G., Falciola, L., Pifferi, V., Visentin, S., D'Incalci, M., Davoli, E.: 3D Mass Spectrometry Imaging Reveals a Very Heterogeneous Drug Distribution in Tumors. Sci. Rep. 6, 1-8 (2016). https://doi.org/10.1038/srep37027

[96] Schulz, S., Becker, M., Groseclose, M.R., Schadt, S., Hopf, C.: Advanced MALDI mass spectrometry imaging in pharmaceutical research and drug development, (2019)

[97] Sun, N., Fernandez, I.E., Wei, M., Wu, Y., Aichler, M., Eickelberg, O., Walch, A.: Pharmacokinetic and pharmacometabolomic study of pirfenidone in normal mouse tissues using high mass resolution MALDI-FTICR-mass spectrometry imaging. Histochem. Cell Biol. 145, 201-211 (2016). https://doi.org/10.1007/s00418-015-1382-7

[98] W. Chumbley, C., L. Reyzer, M., L. Allen, J., A. Marriner, G., E. Via, L., E. Barry, C., M. Caprioli, R.: Absolute Quantitative MALDI Imaging Mass Spectrometry: A Case of Rifampicin in Liver Tissues. Anal. Chem. 88, 2392-2398 (2016). https://doi.org/10.1021/acs.analchem.5b04409

[99] Reid Groseclose, M., Castellino, S.: A Mimetic Tissue Model for the Quantification of Drug Distributions by MALDI Imaging Mass Spectrometry.

Anal. Chem. 85, 10099-10106 (2013). https://doi.org/10.1021/ac400892z

[100] Jadoul, L., Longuespée, R., Noël, A., De Pauw, E.: A spiked tissue-based approach for quantification of phosphatidylcholines in brain section by MALDI mass spectrometry imaging. Anal. Bioanal. Chem. 407, 2095-2106 (2015). https://doi.org/10.1007/s00216-014-8232-7

[101] Tressler, C., Liang, H., Eshleman, J.R., Stumpo, K.A., Glunde, K.: MALDI Mass Spectrometry Imaging of Gemcitabine Treatment in Pancreatic Cancer: Exploring Multiple Matrices to See the Whole Picture. In: Proceedings of the 68th American Society for Mass Spectrometry National Meeting (2020)

[102] Källback, P., Shariatgorji, M., Nilsson, A., Andrén, P.E.: Novel mass spectrometry imaging software assisting labeled normalization and quantitation of drugs and neuropeptides directly in tissue sections. J. Proteomics. 75, 4941-4951 (2012). https://doi.org/10.1016/j.jprot.2012.07.034

[103] Hamm, G., Bonnel, D., Legouffe, R., Pamelard, F., Delbos, J.M., Bouzom, F., Stauber, J.: Quantitative mass spectrometry imaging of propranolol and olanzapine using tissue extinction calculation as normalization factor. J. Proteomics. 75, 4952-4961 (2012). https://doi.org/10.1016/j.jprot.2012.07.035

Chapter 7

Intact Cell Mass Spectrometry for Embryonic Stem Cell Biotyping

Petr Vaňhara, Lukáš Moráň, Lukáš Pečinka, Volodymyr Porokh, Tiziana Pivetta, Sebastiano Masuri, Eladia Maria Peña-Méndez, José Elías Conde González, Aleš Hampl and Josef Havel

Abstract

Stem cells represent a unique cell type that is capable of self-renewal and differentiation into somatic cell types. Since the derivation of human embryonic stem cells and induced pluripotent stem cells, enormous potential has been recognized for disease modeling, drug development and regenerative medicine. Both embryonic stem cells and induced pluripotent stem cells possess the ability to differentiate into all three germ layers, hence they are naturally prone to respond to various differentiation stimuli. These inherent cellular fluctuations, which can result in risky phenotypic instability, must be addressed prior to introduction of these cells to human medicine, since they represent one of the major biosafety obstacles in the development of bio-industrial or clinical-grade stem cell cultures. Therefore, there is an ongoing need for novel robust, feasible and sensitive methods for determination and confirmation of the otherwise identical cells status, as well as for the detection of hidden divergences from their optimal state. A method of choice can be the intact cell mass spectrometry. Here we show how it can be applied in routine quality control of embryonic stem cell cultures.

Keywords: intact cell mass spectrometry, whole cell mass spectrometry, embryonic stem cells, cell culture, culture adaptation, differentiation, quality control

1. Introduction

Embryonic stem cells (ESCs) emerged as an amazing cell biology phenomenon several decades ago. Their derivation represents a milestone in understanding of fundamental molecular and cellular processes during early embryonic development, as well as differentiation mechanisms in somatic cells. Capacity of self-renewal and unlimited differentiation make the human ESCs (hESCs) a promising tool for regenerative medicine, tissue engineering, bio-industry, pharmacological modeling and pollutant testing. However, *in vitro* cultured ESCs suffer from inherent instability, and may develop unwanted properties over time, such as propensity to cancer development or failure of the functional phenotype. Accordingly, the stability of hESCs in long term cultures represents an essential prerequisite for a safe use in medical or industrial fields. Robust, sensitive and feasible tools are therefore required for efficient quality control of ESCs culture.

The beginnings in Mass Spectrometry (MS) development in biology correspond with the discovery of soft ionization techniques, such as Electrospray Ionization (ESI) and Matrix-Assisted Laser Desorption-Ionization (MALDI) by John Fenn (1988) and Koichi Tanaka (1988), respectively, who have been awarded with the Nobel Prize in chemistry in 2002 for these findings. MALDI MS coupled with Time-Of-Flight (TOF) analysis has been then successfully introduced to many fields beyond analytical chemistry, including proteomics, metabolomics, clinical microbiology and structural biology. MALDI TOF MS allows precise identification, structural analysis and quantification of various, predominantly soluble, analytes. MALDI TOF MS can generate molecular or spectral patterns that can be assigned to specific cell types or states, when intact cells are used as the analyte. Moreover it provides sufficient discrimination capacity due to a high variability in molecular species, constituting the whole spectrum a biomarker even without preceding peak identification.

2. Embryonic stem cells and the need of quality control

Shortly after fertilization, early human embryo is composed of blastomeres - equal undifferentiated cells, which have an unlimited potential to develop into any type of embryonic as well as extraembryonic tissue. When the embryo reaches the 8-cell stage, its blastomeres start forming intercellular junctions, maximizing the contact with each other, in a process called compaction. Already in the 16-cell formation, called morula, cells with no contact with the outer environment - the inner cell mass (ICM) - can be recognized. Prior to implantation, morula develops a small cavity, becoming a blastocyst, in which ICM further proliferates and forms embryoblast. Cells of embryoblast, still being pluripotent, differentiate rapidly and build bilaminar germ disc with the distinct layers - hypoblast and epiblast. Epiblast cells represent the essential developmental source for principal embryonic germ layers - ectoderm, mesoderm and endoderm [1].

Mammalian ESCs are derived from undifferentiated ICM of early preimplantation blastocyst. While the embryoblast represents an ephemeral stage in embryonic development, the ESCs, when transferred into long term *in vitro* culture share some characteristics with an immortalized cell line. Specific conditions of the *in vitro* culture prevent differentiation of the ESCs into the embryonic structures, and allow ESCs to keep the molecular machinery necessary for the maintenance of pluripotency and unlimited cell divisions [2].

Development of differentiated structures from embryonic precursor cells is a tightly regulated process. Pluripotent cells, which have not differentiated properly in their developmental time window, or persist ectopically, can form peculiar tumors – teratomas. Teratomas contain regions with haphazard accumulation of mature tissues, such as hair, cartilage, bone, or teeth. Despite the inherent abnormality of teratomas, they were a unique model, which contributed significantly to understanding of embryonic pluripotency and differentiation principles. Indeed, the first cell lines that harbored traits of stem cells were derived from mouse testicular teratocarcinomas [3]. These cells share many features with primordial germ cells, thus they, at the time, served as the first principle model of cell plasticity and embryonic differentiation. Genuine embryonic stem cells were derived from 3.5 dpc mouse blastocyst in 1981 [4], and the long-term culture conditions preventing differentiation were adopted from teratocarcinomas culture protocols [5]. Later, soluble factors critical for maintaining the pluripotency of mouse ESCs (mESCs), e.g. mLIF – mouse leukemia inhibitory factor, and BMP4 – bone morphogenetic factor 4, were identified and allowed mESCs propagation [6, 7].

The embryonic development of primates including humans, however, differs from embryonic development of rodents. That is why the derivation of human embryonic stem cells has not been achieved until seventeen years later by James Thompson [8]. Currently, hundreds of hESC lines and their derivatives are available through curated stem cell repositories and biobanks [9]. A breakthrough in stem cell research came in 2006 when genetic regulators essential for induction of pluripotency were discovered by Shinya Yamanaka [10]. These transcription factors can reprogram fully differentiated somatic cells, so they obtain undifferentiated stem cell-like phenotype. Such human induced pluripotent stem cells (hiPSCs) provide a genuine stem cell model with no ethical burden associated with hESCs [11]. hiPSCs essentially complement the portfolio of cellular models for description of crucial molecular events during embryogenesis, tissue differentiation and cancer development. Nowadays, ESCs derived from embryoblasts and hiPSCs derived by reprogramming of somatic cells represent an important biological model and hold the promise to various clinical applications, including regenerative medicine, tailor made cell therapy and drug testing [12, 13].

The state of pluripotency is developmentally unstable and requires specific molecular machinery to maintain the stem cell phenotype [14, 15]. Developmental trajectories during normal embryogenesis are determined soon, so the complex differentiated structures evolve even in a very early embryo. Sophisticated microenvironment of *in vitro* cell cultures, which is designed and optimized to stimulate unlimited propagation of stem cells and preservation of their full differentiation capability [16], is different from the conditions within the developing embryo. Protocols which represent the golden standard involve additional "feeder" cell layer, which supports cultured stem cells (**Figure 1A**).

Mouse embryonic fibroblasts (MEFs), human preputial fibroblasts, oviduct lining or fetal smooth muscle cells are commonly used as a feeder layer. Feeder cells provide necessary intercellular contacts and produce extracellular matrix (ECM) along with soluble factors, in order to recreate, to certain extent, the microenvironment of a blastocyst [17]. However, co-culture of hESCs with feeder cells bears the risks of carry-over contamination and induction of the immune reaction in a patient being exposed to antigens derived from animal feeder layer upon grafting.

Culture protocols with animal additives are therefore not suitable for the direct use in humans. hESCs culture on defined surfaces coated with ECM mixtures or pure protein layers in complex medium supplemented with essential cytokines

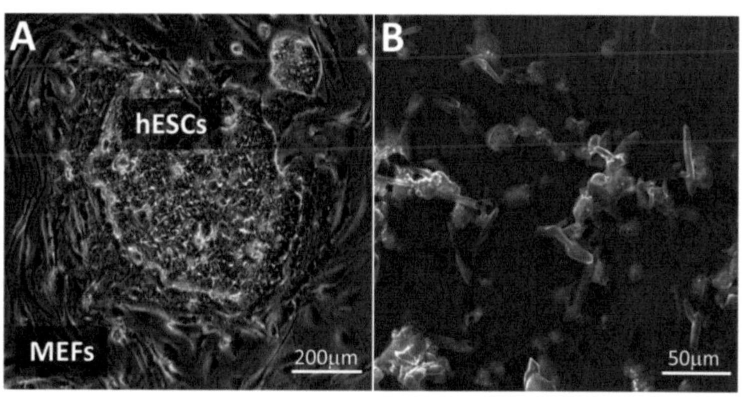

Figure 1.
(A) Single colony of hESCs co-cultured with feeder fibroblast layer (MEFs) as visualized by light microscopy. (B) Scanning electron micrograph of intact hESCs mixed with matrix containing acidified sinapinic acid and spotted on the target plate of the MS instrument.

(e.g. bFGF, activin) may be a solution, if combined with rigorous quality control of other culture parameters (**Table 1**) [18, 19].

2.1 Phenotype shifts and culture adaptation in hESCs

In prolonged culture, pristine hESCs adapt to a two-dimensional, biochemically and structurally less complex, in comparison to a blastocyst, microenvironment [20]. However, the introduction into the *in vitro* culture inevitably induces selection: hESCs clones, which fit the artificial culture conditions the best, are being preferentially selected. Particularly those are clones which divide regularly with optimal doubling time, show reduced sensitivity to programmed cell death, are resistant to passaging method and cell stress arising in culture. Despite stringent culture protocols, such clones can acquire non-random chromosomal aberrations, alterations of epigenetic landscape and changes in gene expression [21–24]. Often, genes which participate in tumor formation or dissemination are involved [25], similarly to those in undifferentiated stem cell lines derived from teratocarcinomas [26].

Darwinian selection in long-term hESC cultures, accordingly, generates cells which remarkably differ from pristine hESCs. Such culture-adapted cells can obtain hazardous phenotype [27], which is similar to a malignant cancer cell line, while keeping normal levels of transcription factors (e.g. Oct 3/4 and Nanog) and other stemness-related molecular markers (e.g. SSEA-4 and Alkaline Phosphatase) [28]. Paradoxically, such significant changes in hESCs, which have acquired the adapted phenotype, can stay unnoticed if cell morphology or expression of stemness factors are not affected. Thus culture-adapted cells can escape routine quality control, which is usually based on visual evaluation or monitoring of a several selected molecular markers. Incompatible cellular alterations are quickly eliminated *in vivo*, typically by the complete rejection of the embryo. *In vitro*, however, aberrant clones may become dominant in culture and disable the safe use of the particular hESCs for clinical or biotechnological applications.

At the moment, there is no routinely applicable method which can reveal hidden shifts in hESCs phenotype or to confirm general stability of a cell culture. The golden standard for authentication of *in vitro* cell lines, e.g. ones derived from individual patients' tumors, is based on analysis of short tandem repeats (STRs). Repetitive sequences dispersed throughout the genome can provide a unique genetic profile and effectively reveal potentially misidentified cell lines [29]. However, STRs analysis is uninformative if used for specific cell culture modifications, such as co-culture of two cell types, analysis of differentiation stages within a single cell line, use of more cell subtypes derived from an individual. Similarly, techniques used for analysis of batch to batch variability, purity of cells, genome or proteome changes, as well as methods focused on a limited panel of biomarkers are not

- microbial and viral contamination
- cell line cross-contamination
- carry-over of animal products in culture
- karyotype instability
- activation of oncogenes and risk of tumorigenicity in patient
- acquisition of immunogenicity and risk of graft rejection in patient
- failure of differentiation into functional phenotype

Table 1.
Risk assessment in clinical grade hESCs culture.

suitable for quality control in routine applications, because they are not informative and often are laborious and costly [30–32].

Successful implementation of pluripotent stem cells into clinical trials requires stringent assurance of the cell product quality and safety, including development of formal methodology for every step of the cells derivation and culture processes. Presence of an unwanted phenotype or deviation from the optimal state of cells in culture may inevitably result in failure of the product and a harm to a patient. Routinely used methods easily reveal microbial and viral contamination, cell misidentifications and chromosomal aberrations which occur in the culture. However, subtle phenotypic shifts or unapparent alterations cannot be detected by microscopy, narrow focused screening for molecular markers or other conventional methods.

Recently, we have been working on the introduction of techniques used in bioanalytical chemistry, chemometrics and in complex data analysis into stem cell field. We have demonstrated that method of choice can be the Intact (Whole) Cell MALDI TOF Mass Spectrometry coupled with sophisticated statistics. Intact Cell MALDI TOF MS is sufficiently robust, sensitive, and cost-effective approach for monitoring of long-term cultures stability and differentiation trajectories of ESCs, therefore it has a potential to complement the portfolio of quality control tools in clinical or bio-industrial applications [33].

3. Intact cell MALDI TOF MS

MALDI TOF MS employs the laser energy to desorb and ionize molecules of an analyte from the crystallized mixture with the matrix, and subsequently separates the resulting ions according to mass-to-charge (*m/z*) ratio. The organic matrix enhances energy transfer to analyte, preserves the structure of the ionized molecules, e.g. peptides, proteins or other biomolecules, and allows their precise structural analysis and identification. In cell biology, MALDI TOF MS is one of the preferred methods for proteomic analysis in a broad range of samples, such as purified or fractioned extracts of cells or tissues. The MS-based proteomics uses protein fragmentation for identification and further generation of a list of unique peptide or protein signatures in wide range of *m/z* values [34]. However, the methodological complexity and the character of the data output may limit the use of traditional proteomics in routine quality control of stem cell cultures, even if coupled with transcriptomics or (meta)genomics.

Even when the intact (whole) cells are used as an analyte, MALDI TOF MS can generate rich spectra without the need of previous cell lysis, fractionation or protein extraction. Mass spectra contain signals for small proteins and peptides, and a variety of other low-mass molecules, including metabolites. Analysis of specific spectral (peak) signatures has been successfully introduced to clinical microbiology, where MALDI TOF MS enables the rapid discrimination, or "biotyping", of bacterial species without the necessity of complex sample processing [35, 36]. Generally the same approach - utilization of relevant spectral patterns as inputs for further processing and analysis [33] - can be used for discrimination of cancer cells [37, 38] or abnormal stem cells in long-term cultures, even in high-throughput setup [39, 40]. Intact Cell MALDI TOF MS was used to identify spectral signatures of glial cells and their classification to astrocyte, microglia and oligodendrocyte type [41]. Principal component analysis then revealed informative peaks for deeper spatial analysis using mass spectrometry imaging in whole brain sections. Similarly, mass spectra have demonstrated to contain sufficient information to reveal the immunophenotype and activation state of immune cells, [42–45] or to classify distinct mammalian cell lines [46, 47]. Moreover, MS can reveal changes associated with molecular phenotype, which occur within cell lines and sublines of common genetic origin. Such approach has been used recently by Povey et al., who demonstrated discrimination of neuroblastoma cell lines sensitive to chemotherapy [48], or by Cadoni et al. who classified ovarian cancer cells sensitive or resistant to cisplatin, based on phospholipid patterns generated by MS [37].

3.1 Intact cell MALDI TOF MS of hESCs

The first step of the preanalytical sample processing is the enzymatic or manual harvesting of hESCs under visual microscopic control. Next, cell clusters are enzymatically disaggregated and washed in isotonic buffers (e.g. phosphate buffered saline, PBS) to remove residual culture medium and additives. PBS has been reported not to interfere with MALDI TOF MS significantly [49]. However, we observed that it may induce random quenching of ionization and decreased intensity of peaks. Therefore, we have added an additional wash with MS fully compatible buffers, such as ammonium acetate [41] or ammonium bicarbonate (ABC) [33], to our protocol, to remove traces of PBS in order to improve mass spectra quality. After cell number assessment, cell are resuspended in 150 mM ABC to desired concentration. Dry cell pellets can be cryostored (at −80°C or lower) with no significant impairment of mass spectra quality.

The MS protocol for hESCs biotyping (fingerprinting) follows the established proteomic or microbiological workflow. Dependent on cell type, instrumentation type and matrix composition, we use typically 1000-25,000 cells per measurement in routine analysis. Cell number can be, though, reduced to several hundred in an optimized experimental design. Cells can be directly placed onto a steel target plate or on transparent indium-tin oxide (ITO) coated glass slides. The ITO coated glass slides enable correlative microscopic analysis in parallel to the MS. In addition, they can be used as a substrate for culture of adherent cells [50].

Sinapinic acid (SA) or α-cyano-4-hydroxycinnamic acid (CHCA) acidified with trifluoroacetic acid are used as a matrix predominantly. SA and CHCA generate uniform-sized crystals, in which cells can be embedded regularly (**Figure 1B**). Although, other matrices, such as 2,5-dihydroxybenzoic acid (DHB) or 2-mercaptobenzothiazole (MBT) can also provide informative output, they form long, needle-like crystals distributed over the target spot unevenly, and therefore are more suitable for solubilized samples.

Routinely, we analyze samples in linear positive mode in *m/z* range of 2-20 kDa, using the usual range of laser energy. Some of the dominant peaks, which have already been partially identified [41, 47, 51], are regularly observed also in hESCs. They correspond to modified histones, thymosin and presumably to ribosomal or other small structural proteins, and can provide an immediate verification of mass spectrum quality.

Processing of the mass spectrum prior to statistical analysis includes reduction of raw data matrix, smoothing of the spectrum, alignment of peaks, baseline subtraction and finally detection of peaks. Average spectrum is then calculated from technical replicates and used to generate a final dataset of *m/z* values with assigned intensities in mV or relative arbitrary units [33, 52, 53].

4. Data analysis

4.1 Mass spectrum as a biomarker

Mass spectrum recorded in a wide range of *m/z* values contains hundreds of charged molecular entities, which together form a spectral profile, or "fingerprint" that can be uniquely assigned to a specific cell type, phenotype or state. However, MALDI TOF mass spectra generated from ionized molecules desorbed from the intact cells are complex and depend strongly on the experimental conditions and preanalytical errors, such as matrix choice, hardware setup and even operator skills. Despite the technical variability, individual mass spectra assembled to a correctly processed dataset may serve as input data for sophisticated mathematical analysis. After the reduction of the unwanted inconsistency, informative patterns in mass spectra can be identified. Finally, processed spectral dataset can be organized in two-dimensional array of cases and intensities of selected peaks. Before statistical analysis is applied to the spectral dataset, preliminary examination of data quality is required. Such rigorous control of data quality includes verification of reproducibility, meticulous calibration and elimination of apparent technical errors or outliers.

Mass spectra of complex biological samples usually contain numerous peaks with rather low intensities and low signal-to-noise ratio. Therefore, the peak detection and recognition is dependent on precise calibration. Where appropriate, we do recommend using the clusters of isotopically pure elements, such as nanoparticles of gold (gold clusters) or black and red phosphorus as calibration standards [54, 55], next to commercially available peptide standards. Mono-isotopic calibrants provide well defined peaks corresponding accurately to predicted mass, allowing proper peaks alignment. Besides, they do not suffer with occasionally problematic indication of isotopic envelope of high mass peptides or proteins.

For evaluation of mass spectra similarity of technical replicates or experimental cohorts, mathematical approaches used in proteomics or metabolomics can be applied. Correlation analysis (e.g. Pearson's correlation, Spearman's correlation, Kendall rank correlation or cosine correlation) can provide a quantitative output that can globally evaluate the similarity of mass spectra [56].

Another relevant factor, which can interfere with the outputs of statistical analysis, is the presence of outlier values in the dataset (case) or within the mass spectrum (peak intensity). Despite the precise laboratory work, outlier values are inevitable and are probably associated with stochastic MALDI effects, as have been already described in bacterial Intact Cell MS [57]. One of the classical procedures, which allows to reveal outliers within the data, is provided by factor analysis and includes careful following of the rank of the data matrix by computing eigenvalues. The number of non-zero eigenvalues (rank of the matrix), visualized in a scree plot, immediately gives the crucial information related to the sample, such as the number of recognizable data groups in mass spectra (e.g. cell types or experimental conditions). In case of two data groups, the rank should equal to two. The presence of outliers is thus indicated by an increased value of the rank [58, 59].

The multivariate evaluation and validation can identify relevant groups or classes within the spectral data. The principal component analysis (PCA) and partial least squares discriminant analysis (PLS-DA) are commonly used. PCA reduces dimensionality of the spectral data, and defines new vectors - principal components, which maximize the variance. Besides, it enables visual observation of the recognized groups of the samples. PCA is an unsupervised method with minimal bias, and its performance is optimal when the intra-group variability is significantly lower than the inter-group variability. It is a well-established tool for processing of complex spectral data, e.g. in proteomics or microbiology [60, 61]. PLS uses different mathematical model than PCA for the distinction of groups. It represents a supervised discriminant analysis, which involves the group information in the algorithm. PLS can provide an excellent discrimination, however, it can suffer from inherent tendency to over-fit the data and identify the clusters even in a uniform spectral dataset. The validation on independent data is therefore recommended [62]. The workflow of data processing in routine analysis of hESCs is summarized in **Figure 2.**

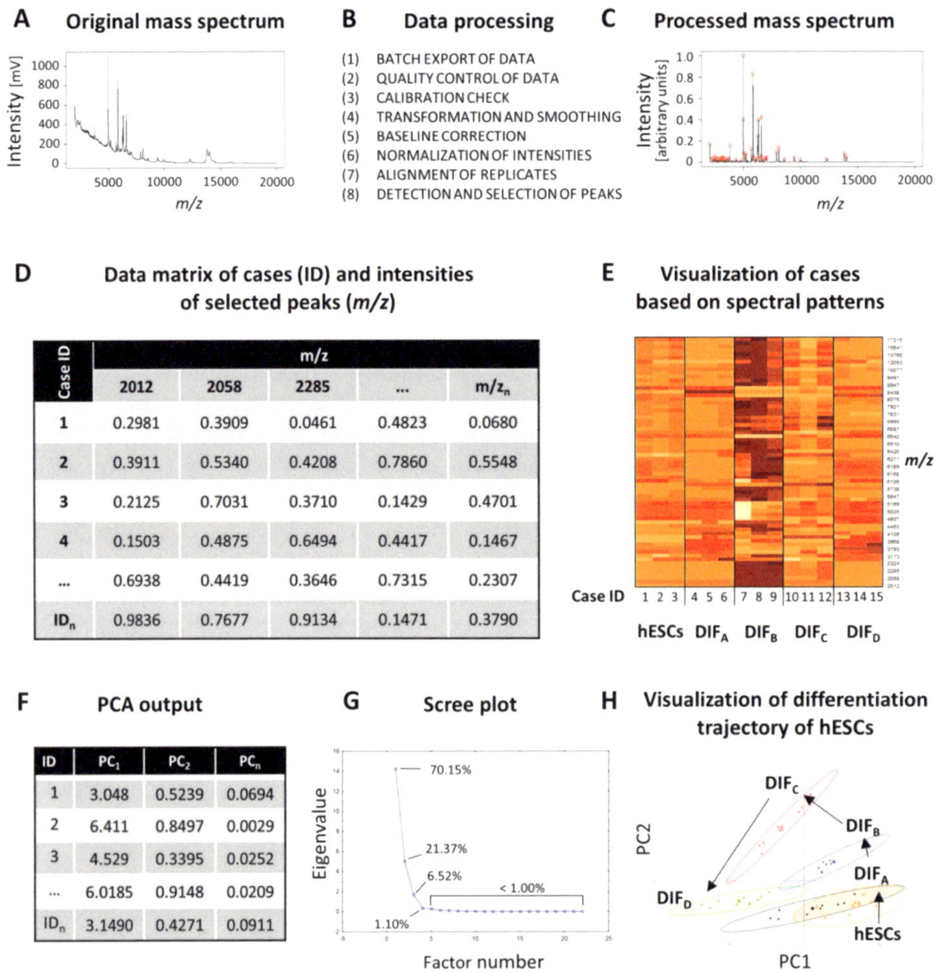

Figure 2.
(A) Example of raw mass spectrum generated from intact cells by MALDI TOF MS, (B) data processing workflow, (C) processed mass spectrum, (D) spectral dataset consisting of individual cases (ID_1-ID_n) and assigned values of peak intensities at defined m/z, *(E) heat map graphically visualizing the dataset containing spectral data of hESCs and four differentiation stages (DIF_{A-D}), (F) example of the output matrix of PCA with recalculated coordinates, (G) scree plot visualizing the number of significant factors contributing to the variability in the dataset, (H) PCA-based visualization of the differentiation trajectory of hESCs progressing towards endodermal phenotype through the four differentiation stages (DIF_{A-D}). Adapted with permission from [63].*

4.2 Classification by machine learning

Artificial Neural Networks (ANNs) represent a non-linear mathematical model, which resembles a brain neural architecture, and possess "learning" and "generalization" abilities. For this reason, ANNs belong to a group of artificial intelligence methods with wide spectrum of complex applications, ranging from purely scientific to industrial or clinical. ANNs utilize diverse types of input data, which are processed in the context of previous training history on a defined sample database to produce a relevant output [64]. The unique chemical fingerprints generated by intact cell mass spectrometry allow the ANN to classify the samples even without preceeding identification of relevant peaks. Successful application of ANNs or any other machine-learning algorithms requires building-up a database of spectral patterns specific for individual cell types, phenotypes or states. This has been successfully achieved in clinical microbiology, however, in eukaryote biology, the

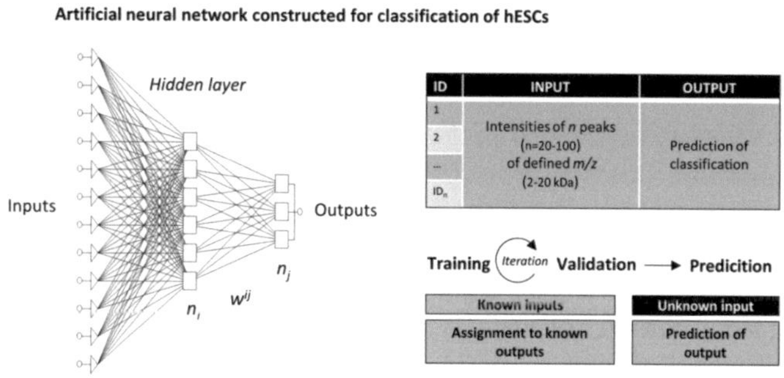

Figure 3.
Architecture of the representative artificial neural network used for prediction of hESC phenotype (output) using peak intensities arranged in a defined spectral matrix (input). Adapted with permission from [63].

complexity of cellular composition and cell plasticity in general represents a major issue. Nevertheless, for the "in-house" databases of well-defined cell models and conditions of their handling and analysis, the Intact Cell MALDI TOF MS coupled with ANNs is a powerful and robust approach that can be easily adapted to any specific application (**Figure 3**).

5. Applications of intact cell mass spectrometry in quality control of embryonic stem cell cultures

Monitoring of clinical-grade stem cells during manipulation, banking or quality control by appropriate tools is the essential prerequisite for their application. We hypothesized that different cell and tissue types or their different states may vary in levels of numerous small molecules, metabolites or peptides and proteins. An unambiguous and unbiased chemical fingerprint obtained by MS can thus reflect such divergences with high sensitivity. In addition, spectral patterns can serve as a highly informative input for subsequent statistical analysis and classification.

To test this hypothesis we used a mouse model of primary hyperoxaluria I - a congenital disorder that affect enzymatic machinery of glyoxylate metabolism. Primary hyperoxaluria I causes oxalate deposits to localize in liver and kidneys, and ultimately lead to hepatorenal failure and extrarenal manifestation of the disease. Alterations of chemical composition within the tissue microenvironment of hyperoxaluric animals can be translated into specific patterns in mass spectra. A dataset, composed of peaks and their corresponding intensities obtained from diseased and healthy animals, was used as an input for cluster and classification analysis and machine learning (ANN) prediction. Spectral patterns clearly distinguished samples from healthy and hyperoxaluric animals and, in parallel, the ANN correctly predicted the category based solely on mass spectrum fingerprint [65].

Generally the same approach can be used for rapid discrimination of cells occurring in stem cell cultures. Mass spectra from pure populations of mESCs, hESCs, and mouse embryonic fibroblasts (MEFs) contain enough information to distinguish the cell types by cluster analysis. Interestingly, these spectral profiles are not lost even in case of mixed populations of two cell types, such as in cross-contaminated cell cultures. Therefore, they can serve as a basis for quantitative estimation of the individual cell types in the mixture. To model such scenario, a broad panel of binary suspension mixtures containing hESCs and MEFs or hESCs and mESCs in defined ratios was prepared. Mass spectra were recorded, processed and the spectral patterns assigned to known quantities of cells in suspension. Resulting dataset then represented calibration data matrix, suitable for quantitative prediction by ANNs. Indeed, well-trained and validated network predicted the number of contaminating cell type with high accuracy in otherwise homogenous and morphologically uniform mixtures. The precision of prediction was dependent on a number of peaks and corresponding peak intensities included in the data matrix. The apparent, dominant peaks with high intensity, did not contribute significantly to the classification. Their exclusion had a negligible effect on the ANN performance, suggesting that there is rather a cumulative effect of minor, but informative peaks, essential for the analysis output [53].

In the long term cultures of hESCs, abnormal clones with the origin in pristine hESCs are inevitably being generated, selected and, finally, expanded. The CCTL 14 hESC line, which has been propagated over hundreds of passages *in vitro* suffers

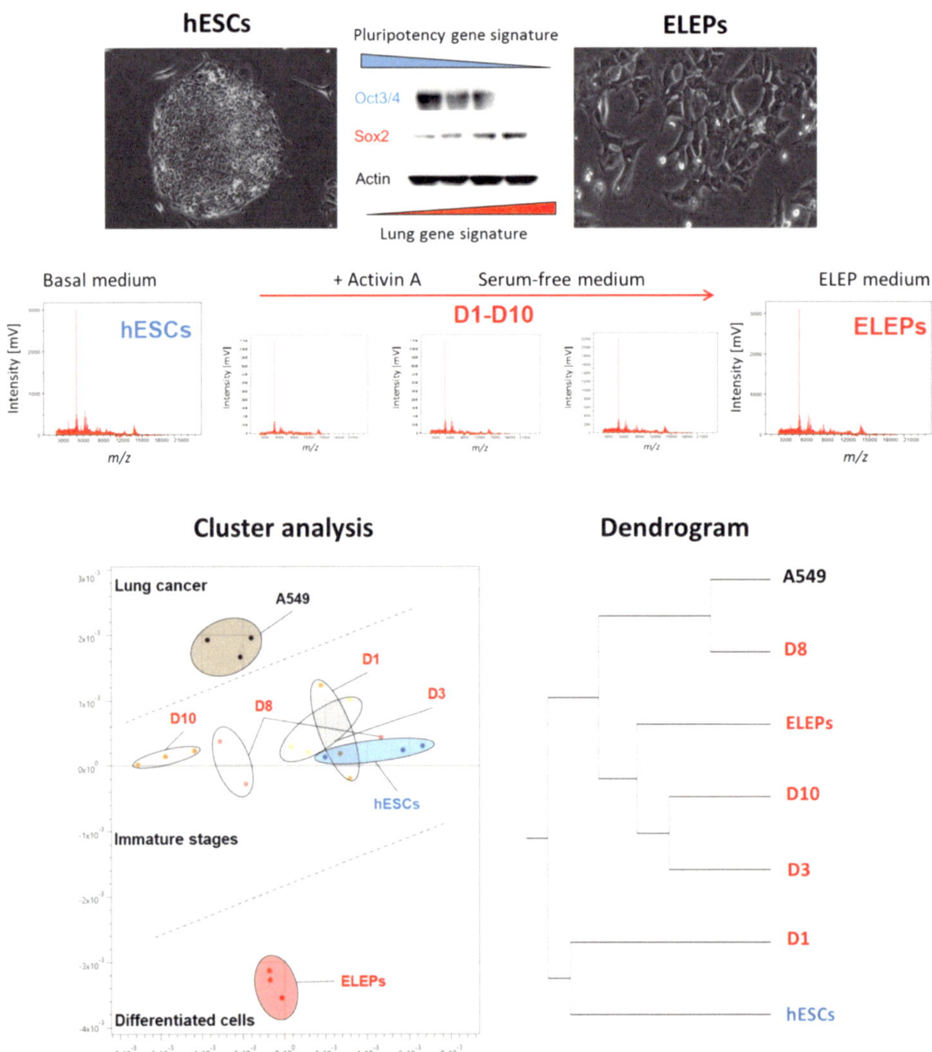

Figure 4.

Monitoring of differentiation of hESCs into early lung progenitors (ELEPs) by intact cell MALDI TOF MS. hESCs induced to ELEP differentiation lose gradually the gene signatures associated with pluripotency and acquire the phenotype of lung progenitors. Differentiation trajectory of cells goes through several immature stages and reaches the final, differentiated stage approximately at day 21. Samples for intact cell MALDI TOF MS were collected prior to induction of differentiation (hESCs) at days 1, 3, 8 and 10 (D1, D3, D8, D10) and finally at the final stage (ELEPs). The lung cancer cell line A549 was used as lung-associated, but abnormal sample class. Cluster analysis, such as PCA or a dendrogram, based only on spectral patterns shows clear separation of classes and indicate that ELEPs represent indeed a distinct cellular entity [66].

from chromosomal abnormalities and shows an increased expression of CD30 and CD44 surface molecules, previously shown to be associated with acquisition of the adapted phenotype and karyotype instability [67]. In addition, adapted, high passage hESCs proliferated with reduced doubling time, and showed lower sensitivity to apoptotic stimuli and decreased capability of differentiation, than pristine hESCs. All the low (P29), mid (P72) and high passage (P269) cells maintained high and constant expression of stemness markers, e.g. Sox-2, and also share the identical morphology [33].

We recorded mass spectra from intact P29, P72 and P269 hESCs cultured under identical conditions, processed and assembled into the dataset containing 255 *m/z* values with assigned peak intensities. The mass spectra showed a high level of visual similarity, however the Pearson's correlation analysis revealed disparities in spectral patterns between P29, P72 and P269 cells. Principal component analysis then correctly discriminated P29-pristine and P269-adapted cell to individual distant cluster, while the P72-transition population cluster has been located and scattered between them. Therefore, such a robust dataset of spectral fingerprints recorded from intact hESCs contains sufficient information to distinguish normal and aberrant cells with otherwise identical morphology and expression of stemness markers. Next, we used the same approach to analyze CCTL 14 hESC line stimulated to differentiation. Generally, cell differentiation means acquisition of novel phenotypic properties, and morphological hallmarks, as well as various molecular mechanisms, involved in cell functioning. Retinoic acid (RA) is a potent inducer of embryonic patterning and neurogenesis *in vivo*, so hESCs induced to the differentiation by RA acquire the neuronal phenotype within several weeks. Mass spectra generated from cells treated with RA for only 24 hrs, contained spectral regions which allowed detection of such early differentiation events even prior to development of clear morphological hallmarks [33].

To test a routine, pre-clinical application, we used Intact Cell MALDI TOF MS to monitor the differentiation of hESCs to early lung progenitors (ELEPs). ELEPs, being the direct precursors of lung cells (pneumocytes), represent an important resource for tissue engineering, pollutant testing and pharmacological analyses. ELEPs gradually differentiate from hESCs under specific culture conditions through several stages (D1-D10). Differentiating immature cells lack an unambiguous marker, which would distinguish them from ELEPs. In addition, during differentiation process aberrant cells with unwanted properties may appear in culture. Intact

+ Feasible, affordable, and robust technique
+ Intuitive, straight-forward approach
+ Unbiased and marker-free classification
+ Discrimination of unapparent, yet critical alterations in stem or progenitor cells, which are not detectable by other techniques
+ Stringent culture protocols and defined SOPs in clinical grade hESCs laboratories allow precise pre-analytical phase
− Inherent instrumental and technical variability of MALDI TOF MS
− Necessity to build own specialized database of spectral patterns
− Calibration sensitivity
− Nontrivial processing and analysis of complex data

Table 2.
Advantages (+) and limits (−) of Intact Cell MALDI TOF MS in quality control of hESCs culture in clinical grade laboratory.

Cell MALDI TOF MS was able to distinguish individual immature stages from hESCs and from ELEPs, as well as from lung cancer cell line (**Figure 4**).

In summary, Intact Cell MALDI TOF MS coupled with advanced statistics provides an efficient tool for revealing aberrant cells in culture or following differentiation trajectories of pluripotent stem cells and progenitors. The advantages and limits of Intact Cell MALDI TOF MS in quality control of clinical grade stem cell cultures are summarized in **Table 2**.

6. Conclusions

Intact Cell MALDI TOF MS reliably discriminates functionally different, but otherwise identical types or subtypes of stem cells of common genetic origin. Moreover, it reveals aberrant or differentiating clones of clinically relevant stem cells or committed tissue progenitors. Coupling the outputs of Intact Cell MALDI TOF MS with sophisticated statistics, such as cluster analysis or machine learning, may provide a feasible and easy-to-use routine tool for quality control of pluripotent stem cells and progenitors long-term cultures.

Acknowledgements

LM is Brno PhD Talent Scholarship holder, funded by the Brno City Municipality, and junior researcher supported by Faculty of Medicine, Masaryk University (ROZV/28/LF/2020) and by Masaryk Memorial Cancer Institute (00209805). Dr. Anna Mac Gillavry Danylevska, Ph. D. is acknowledged for critical reading of the manuscript.

Conflict of interest

The authors declare no conflict of interest.

Funding

This work was supported by Ministry of Health of the Czech Republic (grant no. NV18-08-00299) and by Masaryk University (MUNI/A/1421/2019, MUNI/A/1382/2019).

List of abbreviations

ABC	ammonium bicarbonate
ANN	artificial neural network
CHCA	α-cyano-4-hydroxycinnamic acid
dpc	days post conception
DHB	2,5-dihydroxybenzoic acid
ESI	electrospray ionization
ELEPs	early lung progenitors
ECM	extracellular matrix
hESCs	human embryonic stem cells
hiPSCs	human induced pluripotent stem cells

ICM	inner cell mass
MALDI TOF MS	matrix-assisted laser desorption/ionization time-of-flight mass spectrometry
MBT	2-mercaptobenzothiazole
MEFs	mouse embryonic fibroblasts
mESCs	mouse embryonic stem cells
PCA	principal component analysis
PLS-DA	partial least squares discriminant analysis
SA	sinapinic acid
STR	short tandem repeats.

Author details

Petr Vaňhara[1,2]*, Lukáš Moráň[1,3], Lukáš Pečinka[2,4], Volodymyr Porokh[1], Tiziana Pivetta[5], Sebastiano Masuri[5], Eladia Maria Peña-Méndez[6], José Elías Conde González[6], Aleš Hampl[1,2] and Josef Havel[2,4]

1 Department of Histology and Embryology, Faculty of Medicine, Masaryk University, Brno, Czech Republic

2 International Clinical Research Center, St. Anne's University Hospital, Brno, Czech Republic

3 Research Centre for Applied Molecular Oncology (RECAMO), Masaryk Memorial Cancer Institute, Brno, Czech Republic

4 Department of Chemistry, Faculty of Science, Masaryk University, Brno, Czech Republic

5 Department of Chemical and Geological Sciences, University of Cagliari, Monserrato, CA, Italy

6 Department of Analytical Chemistry, Nutrition and Food Science, Faculty of Chemistry, University of La Laguna, Tenerife, Spain

*Address all correspondence to: pvanhara@med.muni.cz

References

[1] Bardot ES, Hadjantonakis AK. Mouse gastrulation: Coordination of tissue patterning, specification and diversification of cell fate. Mechanisms of Development. 2020;163:103617. DOI: 10.1016/j.mod.2020.103617

[2] De Los Angeles A, Ferrari F, Xi R, Fujiwara Y, Benvenisty N, Deng H, Hochedlinger K, Jaenisch R, Lee S, Leitch HG, Lensch MW, Lujan E, Pei D, Rossant J, Wernig M, Park PJ, Daley GQ. Hallmarks of pluripotency. Nature. 2015;525:469-478. DOI: 10.1038/nature15515

[3] Andrews PW. From teratocarcinomas to embryonic stem cells. Philosophical Transactions of the Royal Society B: Biological Sciences. 2002;357:405-417. DOI: 10.1038/nrg1827

[4] Evans MJ, Kaufman MH. Establishment in culture of pluripotential cells from mouse embryos. Nature. 1981;292:154-156. DOI: 10.1038/292154a0

[5] Martin GR. Isolation of a pluripotent cell line from early mouse embryos cultured in medium conditioned by teratocarcinoma stem cells. Proceedings of the National Academy of Sciences of the United States of America. 1981;78:7634-7638. DOI: 10.1073/pnas.78.12.7634

[6] Bryja V, Bonilla S, Cajanek L, Parish CL, Schwartz CM, Luo Y, Rao MS, Arenas E. An efficient method for the derivation of mouse embryonic stem cells. Stem Cells. 2006;24:844-849. DOI: 10.1634/stemcells.2005-0444

[7] Conner DA. Mouse Embryonic Stem (ES) Cell Culture. Current Protocols in Molecular Biology, 2001;51: 23.3.1-23.3.6. https://doi.org/10.1002/0471142727.mb2303s51

[8] Thomson JA, Itskovitz-Eldor J, Shapiro SS, Waknitz MA, Swiergiel JJ, Marshall VS, Jones JM. Embryonic stem cell lines derived from human blastocysts. Science. 1998;282:1145-1147. DOI: 10.1126/science.282.5391.1145

[9] Holm S. Biobanking human embryonic stem cell lines: policy, ethics and efficiency. Monash Bioethics Review. 2015;33:265-276. DOI: 10.1007/s40592-015-0050-y

[10] Takahashi K, Yamanaka S. Induction of pluripotent stem cells from mouse embryonic and adult fibroblast cultures by defined factors. Cell. 2006;126(4):663-676. DOI: 10.1016/j.cell.2006.07.024

[11] Stroud K, O'Doherty KC. Ethically sustainable governance in the biobanking of eggs and embryos for research. Monash Bioethics Review. 2015;33:277-294. DOI: 10.1007/s40592-015-0047-6

[12] Murry CE, Keller G. Differentiation of embryonic stem cells to clinically relevant populations: lessons from embryonic development. Cell. 2008;132(4):661-680. DOI: 10.1016/j.cell.2008.02.008

[13] Pouton CW, Haynes JM. Embryonic stem cells as a source of models for drug discovery. Nature Reviews Drug Discovery. 2007;6:605-616. DOI: 10.1038/nrd2194

[14] Guenther MG. Transcriptional control of embryonic and induced pluripotent stem cells. Epigenomics. 2011;3:323-43. DOI: 10.2217/epi.11.15

[15] Young RA. Control of the embryonic stem cell state. Cell. 2011;144:940-954. DOI: 10.1016/j.cell.2011.01.032

[16] Odorico JS, Kaufman DS, Thomson JA. Multilineage differentiation from human embryonic

stem cell lines. Stem Cells. 2001;19:193-204. DOI: 10.1634/stemcells.19-3-193

[17] Stacey GN, Cobo F, Nieto A, Talavera P, Healy L, Concha A. The development of 'feeder' cells for the preparation of clinical grade hES cell lines: challenges and solutions. Journal of Biotechnology. 2006;125(4):583-588. DOI: 10.1016/j.jbiotec.2006.03.011

[18] Desai N, Rambhia P, Gishto A. Human embryonic stem cell cultivation: historical perspective and evolution of xeno-free culture systems. Reproductive Biology and Endocrinology. 2015;13:9. DOI: 10.1186/s12958-015-0005-4

[19] International Stem Cell Initiative Consortium, Akopian V, Andrews PW, Beil S, Benvenisty N, Brehm J, Christie M, Ford A, Fox V, Gokhale PJ, Healy L, Holm F, Hovatta O, Knowles BB, Ludwig TE, McKay RD, Miyazaki T, Nakatsuji N, Oh SK, Pera MF, Rossant J, Stacey GN, Suemori H. Comparison of defined culture systems for feeder cell free propagation of human embryonic stem cells. In Vitro Cellular & Developmental Biology – Animal. 2010;46:247-258. DOI: 10.1007/s11626-010-9297-z

[20] Baker DE, Harrison NJ, Maltby E, Smith K, Moore HD, Shaw PJ, Heath PR, Holden H, Andrews PW. Adaptation to culture of human embryonic stem cells and oncogenesis in vivo. Nature Biotechnology. 2007;25:207-215. DOI: 10.1038/nbt1285

[21] Fazeli A, Liew CG, Matin MM, Elliott S, Jeanmeure LF, Wright PC, Moore H, Andrews PW. Altered patterns of differentiation in karyotypically abnormal human embryonic stem cells. The International Journal of Developmental Biology. 2011;55:175-180. DOI: 10.1387/ijdb.103177af

[22] Gokhale PJ, Au-Young JK, Dadi S, Keys DN, Harrison NJ, Jones M, Soneji S, Enver T, Sherlock JK, Andrews PW. Culture adaptation alters transcriptional hierarchies among single human embryonic stem cells reflecting altered patterns of differentiation. PLoS One. 2015;10:e0123467. DOI: 10.1371/journal.pone.0123467

[23] International Stem Cell Initiative, Amps K, Andrews PW, Anyfantis G, Armstrong L, Avery S, Baharvand H, Baker J, Baker D, Munoz MB, Beil S, Benvenisty N, Ben-Yosef D, Biancotti JC, Bosman A, Brena RM, Brison D, Caisander G, Camarasa MV, Chen J, Chiao E, Choi YM, Choo AB, Collins D, Colman A, Crook JM, Daley GQ, Dalton A, De Sousa PA, Denning C, Downie J, Dvorak P, Montgomery KD, Feki A, Ford A, Fox V, Fraga AM, Frumkin T, Ge L, Gokhale PJ, Golan-Lev T, Gourabi H, Gropp M, Lu G, Hampl A, Harron K, Healy L, Herath W, Holm F, Hovatta O, Hyllner J, Inamdar MS, Irwanto AK, Ishii T, Jaconi M, Jin Y, Kimber S, Kiselev S, Knowles BB, Kopper O, Kukharenko V, Kuliev A, Lagarkova MA, Laird PW, Lako M, Laslett AL, Lavon N, Lee DR, Lee JE, Li C, Lim LS, Ludwig TE, Ma Y, Maltby E, Mateizel I, Mayshar Y, Mileikovsky M, Minger SL, Miyazaki T, Moon SY, Moore H, Mummery C, Nagy A, Nakatsuji N, Narwani K, Oh SK, Oh SK, Olson C, Otonkoski T, Pan F, Park IH, Pells S, Pera MF, Pereira LV, Qi O, Raj GS, Reubinoff B, Robins A, Robson P, Rossant J, Salekdeh GH, Schulz TC, Sermon K, Sheik Mohamed J, Shen H, Sherrer E, Sidhu K, Sivarajah S, Skottman H, Spits C, Stacey GN, Strehl R, Strelchenko N, Suemori H, Sun B, Suuronen R, Takahashi K, Tuuri T, Venu P, Verlinsky Y, Ward-van Oostwaard D, Weisenberger DJ, Wu Y, Yamanaka S, Young L, Zhou Q. Screening ethnically diverse human embryonic stem cells identifies a chromosome 20 minimal amplicon conferring growth advantage. Nature Biotechnology. 2011;29:1132-1144.

[24] Wright AJ, Andrews PW. Surface marker antigens in the characterization

of human embryonic stem cells. Stem Cell Research. 2009;3:3-11. DOI: 10.1038/nbt.2051

[25] Ben-David U, Benvenisty N. The tumorigenicity of human embryonic and induced pluripotent stem cells. Nature Reviews Cancer. 2011;11:268-277. DOI: 10.1038/nrc3034

[26] Blum B, Benvenisty N. The tumorigenicity of human embryonic stem cells. Advances in Cancer Research. 2008;100:133-158. DOI: 10.1016/S0065-230X(08)00005-5

[27] Harrison NJ, Barnes J, Jones M, Baker D, Gokhale PJ, Andrews PW. CD30 expression reveals that culture adaptation of human embryonic stem cells can occur through differing routes. Stem Cells. 2009;27:1057-1065. DOI: 10.1002/stem.41

[28] Harrison NJ, Baker D, Andrews PW. Culture adaptation of embryonic stem cells echoes germ cell malignancy. International Journal of Andrology. 2007;30:275-281. DOI: 10.1111/j.1365-2605.2007.00762.x

[29] Almeida JL, Cole KD, Plant AL. Standards for Cell Line Authentication and Beyond. PLoS Biology. 2016;14:e1002476. DOI: 10.1371/journal.pbio.1002476

[30] Kallur T, Blomberg P, Stenfelt S, Tryggvason K, Hovatta O. Quality Assurance in Stem Cell Banking: Emphasis on Embryonic and Induced Pluripotent Stem Cell Banking. Methods in Molecular Biology. 2017;1590:11-16. DOI: 10.1007/978-1-4939-6921-0_2

[31] O'Shea O, Steeg R, Chapman C, Mackintosh P, Stacey GN. Development and implementation of large-scale quality control for the European bank for induced Pluripotent Stem Cells. Stem Cell Research. 2020;45:101773. DOI: 10.1016/j.scr.2020.101773

[32] Wall IB, Davie N. Quality control in cell and tissue engineering. In: Salih V, editor. Standardisation in Cell and Tissue Engineering: Woodhead Publishing; 2013. p. 148-165.

[33] Vanhara P, Kucera L, Prokes L, Jureckova L, Pena-Mendez EM, Havel J, Hampl A. Intact Cell Mass Spectrometry as a Quality Control Tool for Revealing Minute Phenotypic Changes of Cultured Human Embryonic Stem Cells. Stem Cells Translational Medicine. 2018;7:109-114. DOI: 10.1002/sctm.17-0107

[34] Aslam B, Basit M, Nisar MA, Khurshid M, Rasool MH. Proteomics: Technologies and Their Applications. Journal of Chromatographic Science. 2017;55:182-196. DOI: 10.1093/chromsci/bmw167

[35] van Belkum A, Welker M, Pincus D, Charrier JP, Girard V. Matrix-Assisted Laser Desorption Ionization Time-of-Flight Mass Spectrometry in Clinical Microbiology: What Are the Current Issues? Annals of Laboratory Medicine. 2017;37:475-483. DOI: 10.3343/alm.2017.37.6.475

[36] Wang Z, Dunlop K, Long SR, Li L. Mass spectrometric methods for generation of protein mass database used for bacterial identification. Analytical Chemistry. 2002;74:3174-3182. DOI: 10.1021/ac015725f

[37] Cadoni E, Vanhara P, Valletta E, Pinna E, Vascellari S, Caddeo G, Isaia F, Pani A, Havel J, Pivetta T. Mass spectrometric discrimination of phospholipid patterns in cisplatin-resistant and -sensitive cancer cells. Rapid Communications in Mass Spectrometry. 2019;33:97-106. DOI: 10.1002/rcm.8320

[38] Serafim V, Shah A, Puiu M, Andreescu N, Coricovac D, Nosyrev A, Spandidos DA, Tsatsakis AM, Dehelean C, Pinzaru I. Classification of cancer cell lines using matrix-assisted laser desorption/ionization time-of-flight mass spectrometry and statistical analysis. International Journal of Molecular Medicine. 2017;40:1096-1104. DOI: 10.3892/ijmm.2017.3083

[39] Ng EW, Wong MY, Poon TC. Advances in MALDI mass spectrometry in clinical diagnostic applications. Topics in Current Chemistry. 2014;336:139-175. DOI: 10.1007/128_2012_413

[40] Povey JF, O'Malley CJ, Root T, Martin EB, Montague GA, Feary M, Trim C, Lang DA, Alldread R, Racher AJ, Smales CM. Rapid high-throughput characterisation, classification and selection of recombinant mammalian cell line phenotypes using intact cell MALDI-ToF mass spectrometry fingerprinting and PLS-DA modelling. Journal of Biotechnology. 2014;184:84-93. DOI: 10.1016/j.jbiotec.2014.04.028

[41] Hanrieder J, Wicher G, Bergquist J, Andersson M, Fex-Svenningsen A. MALDI mass spectrometry based molecular phenotyping of CNS glial cells for prediction in mammalian brain tissue. Analytical and Bioanalytical Chemistry. 2011;401:135-147. DOI: 10.1007/s00216-011-5043-y

[42] Ouedraogo R, Daumas A, Ghigo E, Capo C, Mege JL, Textoris J. Whole-cell MALDI-TOF MS: a new tool to assess the multifaceted activation of macrophages. Journal of Proteomics. 2012;75:5523-5532. DOI: 10.1016/j.jprot.2012.07.046

[43] Ouedraogo R, Textoris J, Daumas A, Capo C, Mege JL. Whole-cell MALDI-TOF mass spectrometry: a tool for immune cell analysis and characterization. Methods in Molecular Biology. 2013;1061:197-209. DOI: 10.1007/978-1-62703-589-7_12

[44] Ouedraogo R, Textoris J, Gorvel L, Daumas A, Capo C, Mege JL. Analysis and Characterization of Immune Cells and Their Activation Status by Whole-Cell MALDI-TOF Mass Spectrometry. Methods in Molecular Biology. 2019;2024:339-351. DOI: 10.1007/978-1-4939-9597-4_23

[45] Portevin D, Pfluger V, Otieno P, Brunisholz R, Vogel G, Daubenberger C. Quantitative whole-cell MALDI-TOF MS fingerprints distinguishes human monocyte sub-populations activated by distinct microbial ligands. BMC Biotechnology. 2015;15:24. DOI: 10.1186/s12896-015-0140-1

[46] Petukhova VZ, Young AN, Wang J, Wang M, Ladanyi A, Kothari R, Burdette JE, Sanchez LM. Whole Cell MALDI Fingerprinting Is a Robust Tool for Differential Profiling of Two-Component Mammalian Cell Mixtures. Journal of the American Society for Mass Spectrometry. 2019;30:344-354. DOI: 10.1007/s13361-018-2088-6

[47] Zhang X, Scalf M, Berggren TW, Westphall MS, Smith LM. Identification of mammalian cell lines using MALDI-TOF and LC-ESI-MS/MS mass spectrometry. Journal of the American Society for Mass Spectrometry. 2006;17:490-499. DOI: 10.1016/j.jasms.2005.12.007

[48] Povey JF, Saintas E, Aderemi AV, Rothweiler F, Zehner R, Dirks WG, Cinatl J, Jr., Racher AJ, Wass MN, Smales CM, Michaelis M. Intact-Cell MALDI-ToF Mass Spectrometry for the Authentication of Drug-Adapted Cancer Cell Lines. Cells. 2019;8:1194. DOI: 10.3390/cells8101194

[49] Heap RE, Segarra-Fas A, Blain AP, Findlay GM, Trost M. Profiling

embryonic stem cell differentiation by MALDI TOF mass spectrometry: development of a reproducible and robust sample preparation workflow. Analyst. 2019;144:6371-6381. DOI: 10.1039/c9an00771g

[50] Wang S, Chen X, Luan H, Gao D, Lin S, Cai Z, Liu J, Liu H, Jiang Y. Matrix-assisted laser desorption/ionization mass spectrometry imaging of cell cultures for the lipidomic analysis of potential lipid markers in human breast cancer invasion. Rapid Communications in Mass Spectrometry. 2016;30:533-542. DOI: 10.1002/rcm.7466

[51] Kober SL, Meyer-Alert H, Grienitz D, Hollert H, Frohme M. Intact cell mass spectrometry as a rapid and specific tool for the differentiation of toxic effects in cell-based ecotoxicological test systems. Analytical and Bioanalytical Chemistry. 2015;407:7721-7731. DOI: 10.1007/s00216-015-8937-2

[52] Deulofeu M, Kolarova L, Salvado V, Maria Pena-Mendez E, Almasi M, Stork M, Pour L, Boadas-Vaello P, Sevcikova S, Havel J, Vanhara P. Rapid discrimination of multiple myeloma patients by artificial neural networks coupled with mass spectrometry of peripheral blood plasma. Scientific Reports. 2019;9:7975. DOI: 10.1038/s41598-019-44215-1

[53] Valletta E, Kucera L, Prokes L, Amato F, Pivetta T, Hampl A, Havel J, Vanhara P. Multivariate Calibration Approach for Quantitative Determination of Cell-Line Cross Contamination by Intact Cell Mass Spectrometry and Artificial Neural Networks. PLoS One. 2016;11:e0147414. DOI: 10.1371/journal.pone.0147414

[54] Kolarova L, Kucera L, Vanhara P, Hampl A, Havel J. Use of flower-like gold nanoparticles in time-of-flight mass spectrometry. Rapid Communications in Mass Spectrometry. 2015;29:1585-1595. DOI: 10.1002/rcm.7265

[55] Kolarova L, Prokes L, Kucera L, Hampl A, Pena-Mendez E, Vanhara P, Havel J. Clusters of Monoisotopic Elements for Calibration in (TOF) Mass Spectrometry. Journal of the American Society for Mass Spectrometry. 2017;28:419-427. DOI: 10.1007/s13361-016-1567-x

[56] Yilmaz S, Vandermarliere E, Martens L. Methods to Calculate Spectrum Similarity. Methods in Molecular Biology. 2017;1549:75-100. DOI: 10.1007/978-1-4939-6740-7_7

[57] Williams TL, Andrzejewski D, Lay JO, Musser SM. Experimental factors affecting the quality and reproducibility of MALDI TOF mass spectra obtained from whole bacteria cells. Journal of the American Society for Mass Spectrometry. 2003;14:342-351. DOI: 10.1016/S1044-0305(03)00065-5

[58] Brereton RG, Gurden SP, Groves JA. Use of eigenvalues for determining the number of components in window factor analysis of spectroscopic and chromatographic data. Chemometrics and Intelligent Laboratory Systems. 1995;27:73-87. DOI: 10.1016/0169-7439(95)80008-W

[59] Meinicke P, Lingner T, Kaever A, Feussner K, Göbel C, Feussner I, Karlovsky P, Morgenstern B. Metabolite-based clustering and visualization of mass spectrometry data using one-dimensional self-organizing maps. Algorithms for Molecular Biology. 2008;3:9. DOI: 10.1186/1748-7188-3-9

[60] Cejnar P, Kuckova S, Prochazka A, Karamonova L, Svobodova B. Principal component analysis of normalized full spectrum mass spectrometry data in multiMS-toolbox: An effective tool to identify important factors for

classification of different metabolic patterns and bacterial strains. Rapid Communications in Mass Spectrometry. 2018;32:871-881. DOI: 10.1002/rcm.8110

[61] Shao C, Tian Y, Dong Z, Gao J, Gao Y, Jia X, Guo G, Wen X, Jiang C, Zhang X. The Use of Principal Component Analysis in MALDI-TOF MS: a Powerful Tool for Establishing a Mini-optimized Proteomic Profile. American Journal of Biomedical Sciences. 2012;4:85-101. DOI: 10.5099/aj120100085

[62] Worley B, Powers R. Multivariate Analysis in Metabolomics. Current Metabolomics. 2013;1:92-107. DOI: 10.2174/2213235X11301010092

[63] Vaňhara P, Pečinka L, Moráň L, Porokh V, Havel J, Hampl A. Hledání skrytých změn fenotypu – embryonální kmenové buňky na cestě k bezpečné aplikaci. Zpravodaj Česko-slovenské biologické společnosti. 2020;30:8-19.

[64] Amato F, Lopez A, Pena-Mendez EM, Vanhara P, Hampl A, Havel J. Artificial neural networks in medical diagnosis. Journal of Applied Biomedicine. 2013;11:47-58. DOI: 10.2478/v10136-012-0031-x

[65] Houska J, Pena-Mendez EM, Hernandez-Fernaud JR, Salido E, Hampl A, Havel J, Vanhara P. Tissue profiling by nanogold-mediated mass spectrometry and artificial neural networks in the mouse model of human primary hyperoxaluria 1. Journal of Applied Biomedicine. 2014;12:119-125. DOI: 10.1016/j.jab.2013.12.001

[66] Moráň L, Hampl A, Havel J, Kotasová H, Pelková V, Porokh V, Vaňhara P. Intact-Cell mass spectrometry reveals abnormal hESCs in clinical grade stem cell cultures. Conferrence proceedings. The 7th International Stem Cell Meeting. Tel Aviv, Israel, 2019.

[67] Porokh V, Vanhara P, Barta T, Jureckova L, Bohaciakova D, Pospisilova V, Minarikova D, Holubcova Z, Pelkova V, Soucek K, Hampl A. Soluble Cripto-1 Induces Accumulation of Supernumerary Centrosomes and Formation of Aberrant Mitoses in Human Embryonic Stem Cells. Stem Cells and Development. 2018;27:1077-1084. DOI: 10.1089/scd.2018.0017